Discovering Political Ecology

Political ecology is one of the most vibrant fields of environmental research. This book introduces political ecology to a new generation of students in a daring new way: as an interdisciplinary approach to environmental research but also as a series of lived realities and a praxis for change.

The origins of political ecology are often traced through an Anglo-American canon. In *Discovering Political Ecology*, Gustav Cederlöf and Alex Loftus instead take up the challenge of presenting the key conversations and the diverse traditions that have shaped this field with attention to its extensive international roots. Inspired by voices and research in Asia, Africa, Europe and the Americas, the authors address the concerns of those who from different social backgrounds have grown up in a world shaped by climate change and increasing environmental inequalities. Engaging and accessible in style, *Discovering Political Ecology* introduces a set of key concepts around which conversations in political ecology coalesce. It also shows how these concepts allow you to make sense of some of the most pressing issues of our time—issues around water, energy, agriculture, forestry, climate change and environmental justice. Each chapter includes learning outcomes and suggested further readings, extensive bibliographies and seminar activities to be used by students and educators.

At the cutting edge of the field, the book will be of interest to those in all disciplines brought together in political ecology, including but not limited to anthropology, development studies, ecology, geography, history and political science.

Gustav Cederlöf is an Associate Senior Lecturer in Human Ecology at the University of Gothenburg, Sweden.

Alex Loftus is Professor of Political Ecology at King's College London, UK.

This book is a great contribution to political ecology. The examples are chosen from diverse places around the world, yet maintain specificity, striking a good balance between trying to capture a "global" perspective while keeping it locally grounded. It's an interesting and captivating read and I look forward to using it in my teaching.

Alida Cantor, *Assistant Professor, Portland State University Department of Geography, USA*

This is a magnificent field guide to contemporary political ecological thought, informed by multiple scholarly and activist traditions. Written in a direct and accessible style, *Discovering Political Ecology* is the new indispensable text for instructors and researchers alike. It deftly weaves analytical and illustrative points to describe a rich subfield defined by its political commitments, analytical rigor, and growing set of co-conspirators. Political ecology is more important now than ever before, and this book is essential reading for all those who have felt its call to action.

Andrea Marston, *Assistant Professor, Department of Geography, Rutgers University – New Brunswick, USA*

Discovering Political Ecology

Gustav Cederlöf and
Alex Loftus

Routledge
Taylor & Francis Group
LONDON AND NEW YORK

Cover image: © Getty Images

First published 2024
by Routledge
4 Park Square, Milton Park, Abingdon, Oxon OX14 4RN

and by Routledge
605 Third Avenue, New York, NY 10158

Routledge is an imprint of the Taylor & Francis Group, an informa business

British Library Cataloguing-in-Publication Data
A catalogue record for this book is available from the British Library

Library of Congress Cataloging-in-Publication Data
Names: Cederlöf, Gustav, author. | Loftus, Alex, author.
Title: Discovering political ecology / Gustav Cederlöf and Alex Loftus.
Description: Abingdon, Oxon ; New York, NY : Routledge, 2024. |
Includes bibliographical references and index.
Identifiers: LCCN 2023017646 (print) | LCCN 2023017647 (ebook) |
ISBN 9780367558727 (hardback) | ISBN 9780367558734 (paperback) |
ISBN 9781003095521 (ebook)
Subjects: LCSH: Political ecology.
Classification: LCC JA75.8 .C426 2024 (print) | LCC JA75.8 (ebook) |
DDC 304.2--dc23/eng/20230501
LC record available at https://lccn.loc.gov/2023017646
LC ebook record available at https://lccn.loc.gov/2023017647

ISBN: 978-0-367-55872-7 (hbk)
ISBN: 978-0-367-55873-4 (pbk)
ISBN: 978-1-003-09552-1 (ebk)

DOI: 10.4324/9781003095521

Typeset in Bembo Std
by KnowledgeWorks Global Ltd.

Dedication

For Frank, Rosa and Lucian

Contents

List of illustrations

BOX

FIGURES

TABLES

About the authors

Gustav Cederlöf is an Associate Senior Lecturer in Human Ecology at the University of Gothenburg, Sweden. His research is focused on the politics of energy use, infrastructure and urban change, and he has a long experience of working in Cuba and India. Gustav has taught political ecology at Gothenburg, King's College London and the London School of Economics. He is the author of *The Low-Carbon Contradiction: Energy Transition, Geopolitics, and the Infrastructural State in Cuba.*

Alex Loftus is Professor of Political Ecology at King's College London, UK. He has written widely on struggles over water, focusing on the South African city of Durban and, more recently, London. Alex teaches political ecology at King's and has supervised many brilliant PhD students in the field. He is the author of *Everyday Environmentalism*, the co-editor of *Water Politics: Governance, Justice, and the Right to Water*, and Editor of *Antipode: A Radical Journal of Geography*.

Acknowledgements

This is a book about political ecology. Conversations in this field have shaped our understanding of the world since we first encountered it as students. It is our hope that the book will start a conversation in you, stretching the understanding you undoubtedly already have of the inequalities and environmental challenges we face at the current moment.

We are greatly indebted to those who have gone before us and especially to our former supervisors, Raymond Bryant and Erik Swyngedouw. We are also fortunate to work with brilliant students and colleagues in the School of Global Studies at the University of Gothenburg and the Department of Geography at King's College London, who allow us to share our concerns and develop our ideas for change.

In writing the book, we have had excellent support from the team at Routledge, especially Claire Maloney and Andrew Mould.

We dedicate the book to Frank, Rosa and Lucian in the hope that they will live in a world that is not only ecologically sustainable but also more just.

Gustav Cederlöf and Alex Loftus
Gothenburg and London, May 2023

CHAPTER 1

Discovering political ecology

What with my legs aching and my knees hurting,
I'm not giving my cow away. She keeps me fed.

(Svetlana Alexievich, *Chernobyl Prayer*)

One of the most memorable scenes from the miniseries *Chernobyl*, which is based on Alexievich's spine-chilling book, is when a young Red Army soldier evacuates an old woman from her home. A short distance away, reactor 4 at the Chernobyl Nuclear Power Plant has exploded, spewing radioactive materials into the environment, and the woman is busy milking a cow. The conscript tries to convince her she is in danger, but she refuses to leave. She has lived through the Russian Revolution, survived the Holodomor famine under Stalin, and has been asked to leave before, first by Russian and then German soldiers during the Great War. The farm is her home and the cow is her income. "After all I have seen", she says, "so I should leave now, because of something I cannot see at all?" The soldier fears a reprimand unless he evacuates her, and he pours out the woman's bucket with contaminated milk. She still refuses to leave, so he raises a gun. He shoots the cow. "It's time to go", he says.

The drama is a window into a political ecology. The peculiar phrase "political ecology" first appeared in English in the 1970s and gained traction through foundational work in the 1980s and 1990s. Today, it is a fully established, vibrant research field. In the case of *Chernobyl*, it is difficult to distinguish what is "political" and what is "ecological" in the Ukrainian countryside: a long history of war and famine is inscribed in the woman's sense of place, and this history is inseparable from the landscape she inhabits. When seen through the camera lens, it is also a landscape in which social inequality is visibly manifest. The woman leads a modest life in the shadow of an advanced nuclear power station governed by the Soviet state. The plant is located in Chernobyl, but its significance far exceeds that place: it is a national infrastructure purposed for electrification, urbanisation and Soviet socialist progress (Schmid 2015). Investigations into the disaster record that reactor 4 was scheduled for maintenance on the day of the incident. But to compensate for another plant that unexpectedly had come offline, the electrical load dispatcher in Kyiv requested the reactor to stay online. Chernobyl met the energy demands of the Soviet urban population, but as a result, the reactor was left in an exceptionally unsteady state. The benefits of the plant transcended the local area, but the risks associated with it were borne locally.

The ecological transformations following the disaster had deep political implications. Analytically, it is also difficult to understand the local case without a concept of scale, examining

DOI: 10.4324/9781003095521-1

it in relation to wider national and global processes. On the fateful night of the disaster, the prevailing winds over Chernobyl carried the fallout northwest across Belarus and the Baltic. Outside the Soviet Union, the radioactive cloud was first detected in Sweden. Radioisotopes spread with the rain to northern Scandinavia—or Sápmi, as the region is known by the indigenous Sami population. Traditionally, the Sami live from fishing and reindeer herding, but in the year following the disaster, 75 per cent of the produced reindeer meat had to be discarded due to excessive levels of caesium-137. The reindeer fed on lichen which absorbed large amounts of radioactive materials. The effects on Sami livelihoods were devastating in an already politicised environment. Many of the Scandinavian welfare states, which are known for their high levels of social equality, are also settler-colonial projects. For example, about 40 per cent of Sweden's electricity is generated in hydroelectric dams in the country's north, and this low-carbon energy is transmitted to urban centres in the south. But the dams, and the ecological changes brought about during their construction, have a violent history. The Swedish state's claims on the northern rivers in the "national interest" dispossessed the Sami of their right to the rivers, and the damming flooded both reindeer pastures and Sami settlements (Öhman 2016).

To examine the politics of environmental change in Sápmi and Soviet Ukraine, it is necessary to place both cases "within progressively wider and denser contexts" (Vayda 1983: 265), revealing how political and ecological processes inform each other across space and time. Such an approach is integral to political ecology and has developed through work that integrates a range of disciplines, including but not limited to anthropology, development studies, forestry, geography, history, political science, science and technology studies, sociology and systems ecology. However, while an essentially interdisciplinary field, political ecology is often taught through an Anglo-American canon in which its precursor, cultural ecology, is said to have run into a series of problems, necessitating a paradigm shift (Chapters 2 and 3). A series of important texts are then said to have shaped it, with roots in Marxist political economy, peasant studies, feminism and postcolonial studies (Robbins 2020). As one of the earliest and most succinct definitions in the literature has it, political ecology "combines the concerns of ecology and a broadly defined political economy" (Blaikie and Brookfield 1987: 17).

Any political ecologist would acknowledge a far greater diversity in the field than is permitted within this neat origin story, which is also dominated by male voices. Notably, the term *écologie politique* exists in French, though it has only in part developed in dialogue with Anglo-American debates and is strongly associated with environmental politics and activism (Chartier and Rodary 2015; Gautier and Kull 2015). In Spanish, the term *ecología política* has emerged from Latin America, shaped by lived experiences of colonialism and extractivism (Ulloa 2015; Alimonda et al. 2017), but also from Spain where the term bleeds into conversations on ecological economics and environmental justice (Martinez-Alier 2002; Otero et al. 2011). "Other" political ecologies are also debated in studies that to a greater or lesser degree draw on the Anglo-American literature. Scholarship in Portuguese (Freitas and Mozine 2015), the Scandinavian languages (Benjaminsen and Svarstad 2017; Jönsson and Andersson 2017) and increasingly Chinese (Yeh 2015) represents but a few contributions by non-Anglo-Americans that decentre many assumptions at the core of the established canon.

The aim of this book is to consider some of the most critical issues of our time—climate change, deforestation, urbanisation, water insecurity and the imperative of a low-carbon energy transition—from a political ecology perspective. The book also seeks to open the field up to its varied roots and exhibit how the relationships between humans and the environment are studied, written about and transformed by people with often very different historical outlooks reflected in their claims to knowledge. As we will discuss in greater detail, this is not an unproblematic aim: we—the two authors—have both been trained in British universities (if also in Canada, Cuba, South Africa and Sweden), and the book is undoubtedly rooted in the Anglo-American tradition. It provides a situated perspective, albeit one conscious of the importance of making space for a diversity of experiences and research contexts. The book introduces some of the most central concepts and debates in political ecology today, yet our intention is also for it to present the field in a way that helps to destabilise what Soyeun Kim et al. (2012: 36) refer to as "the Anglo-American citadel".

EPISTEMOLOGY: AN INTERDISCIPLINARY APPROACH

Political ecology is a polysemic phrase: in other words, it can mean different things. Conventionally, it is described as an "approach" to environmental research (Table 1.1). As such, it provides a particular outlook on human-environment interactions from which we can gain knowledge about them. At the most basic, two things distinguish political ecology from other approaches: the first is the kinds of questions asked, which are questions that foreground the role of social power, inequalities and conflict in processes of environmental change. The second is the methods employed and the conceptual frameworks used to answer those questions. In part reflecting the field's strong footing in anthropology, political ecologists often employ ethnographic methods sensitive to local specificities and the nuances of lived experience. A rich set of theories has also developed that is put to work to interpret the wider implications of these experiences. The understanding of political ecology as an "approach" is particularly striking in studies from the time the field first emerged in Anglo-American universities; a time when attention to "the political" uncovered questions that earlier research at best had overlooked.

For the rural population of East Kalimantan, Indonesia, the extraction of non-timber forest products, such as rattan (*Calamus* sp.), was long an important livelihood strategy. In an early study, Nancy Lee Peluso (1992) noted that the sources of rattan were quickly being depleted in East Kalimantan. To understand why, she deployed an approach which was "more a method than a theory" (51) and started investigating the role of rattan in the local economy: how was it accessed,

TABLE 1.1 Three ways of understanding "political ecology"

Concept	Focus	Significance	Main Origin
Approach	Epistemology	Particular questions, methods and conceptual frameworks through which environmental issues are studied	Global north
Lived reality	Ontology	Historically produced, geographically located environments in which humans (and nonhumans) are unequally situated	Global south
Practice	Praxis	Practically grounded experiences leading to theoretical insights and theoretically informed action for change	Global north and south

produced and protected? She then increased the level of abstraction and positioned the local land users in the context of the national and international political economy. Over the past decades, she argued, the Indonesian forestry sector had seen a commercial boom, which altered the social and economic organisation of the rattan trade. The integration of postcolonial Indonesia into globalised trade networks commodified the forest and rapidly incentivised extraction. The lure of economic expansion ultimately destroyed rattan habitats, degraded the forest ecosystem and undermined local livelihoods. Peluso's approach was to start in the local context and successively work her way outwards in space and backwards in time to examine the politics of environmental change.

Peluso was part of a first generation of political ecologists who refined the approach to explain issues ranging from deforestation to famine and land degradation. Land degradation refers to a loss in the productive capacity of the land, and advocates of a theory known as neo-Malthusianism long argued that it results from excessive population pressure and overgrazing (Chapters 4 and 8). In a prominent study, Piers Blaikie (1985) exposed the limits of such an apolitical mode of explanation. It is not the sheer number of people and animals that degrade landscapes, he argued, but rather, local land users are integrated into wider nets of social relations that influence their decisions. For example, Blaikie found that farming families incorporated in a market economy had to produce more than they consumed to make a profit. Controverting neo-Malthusian explanations, he concluded that when "surpluses are extracted from cultivators who then in turn are forced to extract 'surpluses' ... from the environment ... [this in time] leads to degradation" (124). Land degradation, then, was above all a political economic phenomenon.

The confluence of ecology with political economy was paramount. Scholars like Peluso and Blaikie, versed in Marxist political economy, drew attention to the relationship between society and the environment and questioned how the mediating role of social power ought to be understood in that relation. The political ecology approach converged on two key concepts: first, careful attention to social power and how power relations influence the practices of people in a local context; and second, the implications of scale—that is, how the local articulates regional, national and global processes. However, scholars refining the approach also initiated a debate that remains unresolved to this day. By tracing some of the key moments in this debate, we can see how the development of political ecology conventionally is narrated in the Anglo-American tradition.

In the late 1990s, Raymond Bryant and Sinéad Bailey (1997) formulated a set of shared assumptions which, they argued, formed the basis of the political ecology approach. In doing so, they also argued for the need for "putting politics first" in research on environmental issues—at the expense of ecological processes. They wrote:

> the relationship between politics and ecology is not an equal one. In effect, the role of politics in shaping ecology is much greater today than in the past as a result of rapid social and technological changes that render problematic the idea of a "natural" environment ... and may soon bring under human control the operation of ecological processes themselves. (5)

The argument resonates with contemporary discussions of "the Anthropocene", a concept suggesting that human activities now are the dominant influence on climatic and environmental processes. While many political ecologists rightly remain sceptical about the Anthropocene concept (Chapters 3 and 4), it is widely established that "nature" cannot be seen as a mere canvas for human life. Rather, nature is transformed by human activities and is, at least in part, socially produced. This insight underscores Bryant and Bailey's argument about the primacy of politics: human activities must be foregrounded because "to understand, as well as to be in a position to solve, the … environmental crisis is to appreciate the ways in which the status quo is an outcome of political interests and struggles" (5).

The reaction to Bryant and Bailey's intervention was immediate. In a vociferous critique titled "Against political ecology", Andrew Vayda and Bradley Walters (1999) argued that by assuming that larger-scale social processes influence how people engage with the environment, there is a risk of overlooking the often-complex ecological interactions through which environmental changes occur. It was presumptuous to insist on the importance of political economy. In response, Susan Paulson et al. (2003) found that Vayda and Walters made a problematic distinction between political and ecological explanations. Instead, environmental events can never be seen as either biophysical or social—they are always both-and. Michael Watts and Richard Peet (2004) held Vayda and Walters to account for their claim to be working without guiding assumptions when they searched for explanations of environmental change. To gain knowledge about a phenomenon, it is necessary to define and delimit it, and Vayda and Walters failed to explain *why* some causes should be considered and others be left out. By emphasising perspectives rooted in political economy, Watts and Peet argued, political ecologists accounted for the assumptions and the contingent choices they inevitably made. Thus, their research became more transparent, and they were able to systematically revise the theoretical assumptions with which they worked.

The controversy was ultimately overturned by wider changes in the field. With a careful reading of the literature, the claim that political ecologists showed little interest in ecology was curious to begin with (Walker 2005). But it also missed the point of much recent work. Around the year 2000, feminist scholars were drawing attention to how women often experience environmental changes differently than men, and they showed how the genders, as categories of cultural difference, were reproduced in socially produced environments (Rocheleau et al. 1996; Harcourt and Nelson 2015). Urban scholars, in turn, questioned ingrained ideas about the city as a domesticated space, distinct from "the rural" and "the wild". Just like feminist political ecology, urban political ecology developed as a subfield, approaching the city as a form of produced nature in which life is unthinkable without the uneven flows of water, energy, food, waste and sewage that "metabolise" it (Figure 1.1; Heynen et al. 2006). The scholarship with the most significant impact on the field, however, was employing social constructivist perspectives. In studies of environmental change, what did "the environment" mean? What constituted "nature" as an object of inquiry? And vitally, *who* had the right to define that which passed for "nature"?

Poststructuralist and postcolonial scholarship brought a new understanding of human-environment relations to the fold. As Arturo Escobar (1999: 1) argued, studies focusing on how discourses and knowledge are interlaced with power (Chapter 2) signalled "the final decline of the modern ideology of naturalism, that is, the belief in the existence of pristine Nature outside of history and human context". In recent studies, such as Yaffa Truelove's (2019) work on unequal water access arrangements or Seema Arora-Jonsson et al.'s (2021) research on violence and care in forest management, it is evident that epistemological questions—that is, how we gain knowledge about the environment—still take centre stage in the field. The crucial insight is that the ways in which "nature" is described are not natural or value-free. Rather, they reflect the situated interests of particular groups while marginalising the voices and knowledges of others.

FIGURE 1.1 Urban nature? An *organopónico* urban farm in Cuba

In most modern cities, urban consumers access food through globalised trade networks that connect them with rural producers, near and far. While city life is unthinkable without a sustained supply of food, the dislocation of food production from everyday urban experiences has generated particular forms of urban nature (Chapter 5). Urban agriculture, by contrast, reconfigures the socio-ecological relations that "metabolise" the city. For example, since the 1990s, organic urban agriculture has proliferated in Cuba and has become an integral part of the country's otherwise ailing state-economy (Cederlöf 2016). It has also produced radically different urban political ecologies compared to those based on globalised commodity trade.

Photo by Gustav Cederlöf

At the core of these discussions is the question *who*. Who can access particular resources? Who benefits from environmental change? And who carries the costs? The insistence on asking the *who*-question is one of the field's distinguishing features. But crucially, social constructivist perspectives revamped the *who*-question. Studies emphasising political economy often assume that environmental actors exist *a priori* and come into conflict with each other in environmental struggles (although we should stress that not all studies with this focus do so, as notably is evident from Watts' 2013 classic book *Silent Violence*). Building on a Foucauldian notion of power, however, it was also pertinent to ask how people come to identify as particular subjects in a given setting—to study how subjectivities emerge historically. One of the most exciting areas of research today, which we discuss in detail in Chapter 8, builds on conversations around "environmentality". This scholarship examines the mechanisms through which people come to identify as subjects via their interactions with the environment in a particular institutional setting (Agrawal 2005; Li 2007; Valdivia 2015).

The debate on how the relationship between politics and the environment is best conceptualised is still unfolding. While the arguments stressing political economy and social construction are indelible, familiar questions have recently been asked again: what about the implications of ecology, climate systems, wetlands, wildfires and zoonotic viruses (like COVID-19) in human life? And what about the roles of materials, plants and animals in environmental struggles? The renewed interest in ecology partly stems from the political questions physical geographers and field ecologists are asking in relation to their work (Lave et al. 2018; Baker et al. 2019), but it also reflects three evolving research agendas. First, a series of studies has recently asked how the biophysical and chemical properties of materials and resources "matter" in socio-economic life (Barry 2013; Bakker and Bridge 2021). In a study of water, for instance, Karen Bakker (2012) makes the seemingly straightforward observation that water is a non-substitutable resource for human life and ecosystem function. But she also notes that the constant need for water is counteracted by its variable "flow" in time and space. This variability physically conditions attempts to regulate water resources. For Bakker, it is, therefore, necessary to investigate how the material flow of water is imbricated in paradigms of water governance and the production of knowledge about this resource. The question is not so much what water *is* but what it *does* in the context of social, economic and political life.

Second, through work at the interface of political ecology and ecological economics, the material aspects of the political economy are receiving new attention. Ecological economics first developed out of dissatisfaction with the way mainstream economics treated the environment as an "externality" to the circulation of economic value. Ecological economists instead conceptualise the economy as a system of material and energy flows, which is an integral part of value creation (Spash 2017). The materialist perspective is raising novel questions for political ecologists: what institutions, social relations and cultural constructs sustain the flows of materials through the globalised economy, and how do those flows reproduce social inequalities (Otero et al. 2011; Newell et al. 2017; Givens et al. 2019)? What roles do ecological concepts, such as the laws of thermodynamics, play in our understanding of socio-economic development (Cederlöf 2021)? And indeed, one of the most contentious conversations today derives from the convergence of political ecology and ecological economics: advocates of "degrowth" argue that an economy dependent on exponential growth, by definition, is unsustainable on a planet with a finite resource base. The degrowth hypothesis holds that to achieve a just climate transition, it is necessary not only to reduce the levels of material and energy throughput in the global economy but also to develop cultural concepts other than "growth" to guide social development (Chapter 7; D'Alisa et al. 2014; Kallis et al. 2020).

Finally, deeply seated ideas about the human subject as the central focus in political ecology are currently being challenged by so-called "more-than-human" or "posthumanist" scholarship. In its various iterations, the *who*-question has been reserved for humans in the field, yet more-than-human research radically decentres the human subject as the protagonist of history. In a remarkable study, Anna Tsing (2015) takes the perspective of the matsutake mushroom (*Tricholoma matsutake*), which is prized as one of the most exclusive gifts in Japanese culture.

Matsutake only grows in pine forests managed by humans since broadleaf regrowth otherwise shades out the mushroom's host pines. Tsing traces the connections that exist between pines and fungi, a socially marginalised labour force, globalised trade networks and the history of Japanese national identity to ascertain how species are co-dependent on each other. Moreover, she shows how nonhuman species, despite attempts at domination, act in highly uncontrollable ways. The socially unequal matsutake economy can only be understood from a relational, multispecies perspective in which mushrooms have as much agency as human beings. In Chapter 4, we explore why these more-than-human arguments remain fiercely contested in the field, for example for their tendency to obscure power relations and for reproducing indigenous epistemic traditions without recognising these (Lave 2015; Davis et al. 2019; Durand and Sundberg 2022).

ONTOLOGY: A LIVED ENVIRONMENT

The main protagonists in our narrative so far have been Western scholars whose concerns about environmental degradation and social inequality have guided the development of the field. Writing in the Brazilian journal *Desenvolvimento e Meio Ambiente* (Development and Environment), Enrique Leff (2015: 34) makes an astute observation: "Establishing the field of political ecology in the geography of knowledge is a … complex endeavor", he writes. Leff points out that political ecology has had a different impetus in the global south. In Latin America, for example, where European colonisation classified non-European bodies as racially inferior and at best marginalised indigenous knowledges (Chapter 2; Grosfoguel 2013), political ecology has emerged "from a politics of difference rooted in the ecological and cultural conditions of its peoples" (34). The motivation to discuss environmental issues in terms of "political ecology" is thus part of a collective struggle for emancipation and justice in an environment produced during a history of colonisation and decolonisation.

When political ecology is described as an "approach" to environmental research, the human observer is placed outside, and in relation to, the environment under study. However, if we start narrating political ecology through the work of scholars like Leff, rather than for example Peluso and Blaikie, it is evident that political ecology should be described less in terms of epistemology—that is, how we gain knowledge about and explain environmental issues—and more in terms of ontology: about what the environment *is* and *for whom*. This is a qualitatively different way of understanding the phrase "political ecology". In Sápmi and Soviet Ukraine, we saw how landscapes are the products of contentious historical trajectories, geographical connections, unequal power relations, economic activities and cultural perceptions. Since people have different embodied experiences of these landscapes—whether they are small-farmers or urban residents, reindeer herders or a state-owned hydroelectric utility—the "environment" must be seen to *be* a political ecology. "[W]hat is suggested here", as Bengt Karlsson (2015: 354) writes, "is the impossibility to think of the environment as external to the social; instead one must engage with the co-constitution of nature and society".

Next to power (Chapter 2) and scale (Chapter 3), as we have discussed above, "nature" is a keyword for political ecologists. If humans inhabit political ecologies, what nature *is* and how we gain knowledge about it are cardinal questions. In Chapter 4, we introduce contributions that show how nature is far from the taken-for-granted "nonhuman" it is often seen to be and instead can be thought of as "constructed" or "produced". A closely connected concept is "urbanisation", which is having an increasing impact on political ecological debates (Chapter 5). Urbanisation is the process through which cities form, and while urbanisation gives rise to distinctive urban natures, or ecologies *in* cities (Nagendra 2016; Gandy 2022), urban life is inconceivable without considering the ecology of the city. This refers to the material flows and the often-distant land uses that enable the urban experience. Urbanisation is a process that not only produces the city but also the rural hinterlands that sustain urban life. As Bruce Braun (2006: 218–219) argued some time ago, all nature is in some sense urban nature today:

> for to the extent that systems of production, exchange and consumption have become global, "distant" natures and everyday urban environments are woven into tight webs of socio-ecological and spatial relations. … Add in the consumption practices of the city's residents, and these political-ecological networks extend to the ends of the earth, taking in ecological and labour conditions from the coffee plantations in Brazil to the sweatshops of mainland China.

Dominated by Western voices, urban political ecology has had a strong focus on cities in the global north, but the argument for urbanisation as a planetary condition resonates with a critique of the globalised economy emanating from the south. While urbanisation, industrialisation and "development" have engendered particular experiences in the global north, the same processes have had qualitatively different effects in other parts of the world. The development idea first emerged during the Enlightenment to suggest that all of humanity shares a common history; a history in which societies begin at a simple, primitive stage and through time become more complex, modern and urban. During the colonial period, Europeans envisaged the human species to be divided into variant "races", each of which had reached a different developmental stage. Europeans had reached the pinnacle of modernity, and there was a belief that other races could see their own future for example in British, Dutch and French history (Said 1979; Metcalf 1995). In the postcolonial period, different nations are instead conventionally categorised according to their level of "development", aspiring to replicate an idealised Euro-American experience (Escobar 1995).

Within this Western cultural paradigm, the overarching goal of government is to achieve "development". The pursuit of development is, in turn, closely linked to a vision of the human subject as external to the environment, a subject using "nature" as a resource to its own benefit. Arguing against such a dualistic (or Cartesian) worldview, Latin American political ecologists have posited that Western modernity is unthinkable without a complementary concept of "coloniality" (Chapter 2; Quijano 2000). Coloniality identifies the subordinate role of formerly colonised populations in the project of modernisation, supplying labour and resources to urbanised, more "developed" regions of the world from a global periphery. As Eduardo Gudynas (2016: 724) argues, "modernity should not be considered as the imposition of European culture since the Enlightenment, but as a more complex phenomenon shaping the world-system, produced in Europe but also in the colonies at the same time".

From a perspective like this, it is evident that the goals of human life and the relationships between humans and nature can be conceptualised in radically different ways from the Western mainstream. Gudynas, for example, draws attention to the Andean concept of *buen vivir* (approx.

"good living"), which is rooted in the Aymara notion of *suma qamaña* (in Bolivia) and the Quechua expression *sumak kawsay* (in Ecuador). *Buen vivir*, Gudynas argues, is not a "variety of development" on a par with Western neoliberalism or Chinese state-capitalism, which by different means strive towards the same "development" outcome. Rather, it is an alternative to development altogether. *Buen vivir* is a cosmology that rejects dualisms between humans and nonhumans and instead understands nonhumans to be part of social communities. Attention to indigenous cosmologies and what Escobar (1995) calls "post-development" has generated provocative discussions among political ecologists. It is regularly assumed that one material reality exists over which different social groups struggle for interpretation (which is a key insight of poststructuralist scholarship). But proponents of the "ontological turn" argue that multiple realities, reflecting different ontological assumptions, must be seen to exist in parallel. Multiple realities are thrown together in a struggle for dominance in a "universe" or co-existence in a "pluriverse" (Figure 1.2; Blaser 2014; Kothari et al. 2019).

The argument for a pluriverse raises unresolved questions of relativism, but Astrid Ulloa (2015) stresses how cosmologies like *buen vivir* attest to the possibility of practising human-environment relations differently at a time of global environmental crisis. Alternatives to development accentuate that there is an outside to the dominant development paradigm. This is most obvious when the argument is voiced *from* this constitutive outside: "As an Anishnaabe and Haudenosaunee woman", Vanessa Watts (2013: 22) writes, "my worldview is continuously tested against the colonial frame". Thus, we can examine how the Western development model can be reconfigured in terms of "sustainable development" (Adams 2020), but more radically, as Ulloa argues, we can also inquire "how local groups establish a political space to act in a collective way, in order to generate their own relationships between humans and with nature" (329). How are local groups producing alternative political ecologies? Ulloa asserts that territorial autonomy and environmental self-government are two necessary enabling factors for the

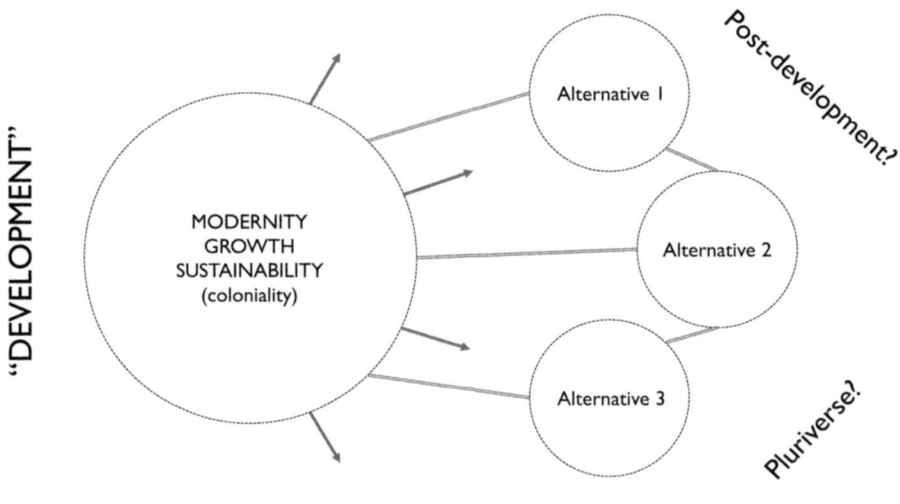

FIGURE 1.2 Whose worldview counts? From the Western development paradigm to a "pluriverse" of lived realities

emancipation of indigenous communities in the pursuit of decolonisation and environmental justice. Political ecology, then, can be described as an approach to environmental research, but it is also a lived reality that exists and can be transformed through practice.

PRAXIS: GAINING AND GIVING VOICE

Political ecology, as we are arguing, is a phrase with multiple meanings. A third way of thinking about it is as a practice: that is, as something people *do*. "[P]raxis", Farhana Sultana (2021: 158) writes, "is generally understood to be theory in action, or practice and theory that inform each other". Indeed, it is by participating in different political ecologies, and doing so in a reflexive manner, that we generate theoretical insights. But we can also expose inequalities and act for change by putting the theoretical insights we make into practice. Most political ecologists are active in universities, as lecturers, researchers and postgraduate or undergraduate students, and see education, activism and policy-relevant research as their main pathways to change (Figure 1.3).

FIGURE 1.3 The Political Ecology Network—POLLEN

POLLEN was founded in 2015 to connect political ecologists globally and be a platform for discussion and collaboration. The map shows the "nodes", which are member university departments, that form the network. There are, of course, many political ecologists who are not members of POLLEN, yet the map suggests that the geography of where political ecological knowledge is produced is very particular: it is centred on Europe (and especially the United Kingdom) and the United States. Location matters as the research contexts within which knowledge is generated often also set the parameters defining that which counts as important and prioritised areas of study.

Source: http://politicalecologynetwork.org/pollen-nodes

Yet the ability to communicate and act on political ecological issues far transcends the academic arena. The political ecology community incorporates artists, development practitioners, environmental activists, journalists, policymakers and citizens on all continents who are concerned about the status quo (Batterbury 2015). This book aims to introduce key concepts and themes in political ecology, but we also hope to embolden you to act for a more just sustainable world.

As a practice, political ecology has evolved in positive directions in recent years. Reflecting on his long-term involvement in the field, Bryant (2015a) notes that a more heterogenous range of voices is heard today, even as the ambition to promote subaltern perspectives always has characterised the field. Changing praxis, he writes, is reflected in "the growing number and prominence of women … [and] greater inclusivity in terms of scholars who are of non-European (or in some cases partly non-European) heritage, writing and teaching in political ecology" (18). As more people, and increasingly nonhumans, gain voice, new conversations are started. Changing praxis is, therefore, also a question of giving voice, when people in privileged positions make space for experiences that redefine how political ecologies are studied and acted on: "The point", as Kim et al. (2012: 36) write, "is to recognize that perceptions of political ecology inevitably reflect who is doing the perceiving and where these observers are doing the perceiving from".

Our ambition is to contribute to an ongoing conversation that identifies a need to "facilitate a more differentiated gaze" (Kim et al. 2012: 35), "giv[e] voice to a wider set of research concerns" (Bryant 2015a: 20) and "decentre" the Anglo-American canon (Loftus 2019: 174). In the writing process, we have come across two challenges that speak directly to the complexity of the issue. First, education is often seen to be a part of the "solution" to environmental issues and the climate emergency. Yet education can equally reproduce unsustainable human-environment relations and a skewed representation of voices depending on the curriculum being taught and the pedagogy being used (Kahn 2010; Daigle and Sundberg 2017; Hjorth Warlenius 2022). As noted, the two of us writing this book have both been trained in the Anglo-American tradition. Both male, we were also supervised during our PhD studies by other male political ecologists whose work is often positioned at the forefront of this tradition. Through our writing, we have realised how deeply ingrained the established narrative of the field's development is in us and how difficult it is to "undiscipline" your own thinking. The geographies of knowledge represented in this book are unquestionably limited by our frames of reference and our attempts to negotiate with the privileged tradition we are a part of. In spite of the ambition to reference broadly and inclusively, it is never possible to be fully comprehensive. Writing involves a process of selection and is inevitably partial in perspective.

Second, the production of knowledge always takes place in an institutional context that defines the parameters of that which can be said. When the publisher first approached us to ask if we were interested in writing a new political ecology textbook, the question likely came to us (as opposed to someone else) because of the networks having formed over the years between Routledge and scholars at the university where we both worked at the time (King's College London). The next step in the publishing process is to write a proposal indicating what the book will be about. Here, we were very open with our ambition to write an alternative introduction to the field—"to introduce a quasi-canonical political ecology", as we put it. However, in part reflecting the peer-review reports on the proposal, in part reflecting the publisher's commercial interests, we were asked to also narrate the field based on the established Anglo-American canon so that the book could "stand alone" as an "introduction to the field". As you will have noticed, we have already started doing so in this chapter. The issues of representation, recognition and "giving voice" are deeply ingrained in the institutions of research, education and publishing. Navigating this terrain, we hope to have found a balance between the "need" to anchor the book in the established canon—dominated as it is by predominantly male Anglo-American scholars—and our desire to present the field in a way that is more open to its diverse roots.

CONCLUSION AND OUTLINE OF THE BOOK

So, what is political ecology? In this book, we argue that political ecology can be understood in three complementary ways. First, it is *an interdisciplinary approach* to the study of environmental issues, and we introduce a set of conceptual tools that political ecologists use to examine how social inequalities, economic activities and political processes are interwoven with environmental change. Second, it describes the environments humans inhabit in socially unequal ways—*lived realities*. Throughout the book, we explore a range of political ecologies shaped by complex historical trajectories and interconnected global processes. Finally, it is *a practice* through which human-environment relations are enacted and transformed. In this sense, political ecologists critically examine dominant practices but also reimagine and build on alternative human-environment relations to overcome the impasse at this time of environmental and climate crisis.

In the first four chapters that follow this introduction, we present key concepts that all lie at the heart of political ecological inquiry: power, scale, nature and urbanisation. The first two are particularly salient when political ecology is seen as an approach to environmental research. The latter two speak more directly to an ontological understanding of political ecology. Another four chapters then follow which each takes one of the most topical themes in the field as its starting point. Here, we demonstrate how political ecologists apply their conceptual tools in relation to questions of water, energy, fields and forests, as well as the "virtual", which refers to the resources and emissions embedded in everyday commodities. Each chapter ends with a concluding summary and a list of learning outcomes that distil the key insights. We also recommend further readings, reflecting the chapter theme. At the end of the book, a final section provides a learning activity for each preceding chapter. We hope that these activities will allow you to deepen your knowledge, in class or individually, and start putting political ecological perspectives into practice.

FURTHER READING

The political ecology literature is expansive, and beyond this book, several English-language texts have been established as common points of reference in the field. Paul Robbins' *Political Ecology* (2020), now in its third edition, has long been recognised as an effective textbook. Robbins is often cited for his claim that political ecology is both "a hatchet" and "a seed", providing the tools for critical analysis but also radical action. Bill Adams' *Green Development* (2020), now in its fourth edition, focuses on the discourse and practice of "sustainable development" and, in doing so, introduces several key topics in political ecology. The sections on nature conservation offer an excellent complement to this book. The conservation focus also characterises Tor Benjaminsen and Hanne Svarstad's *Political Ecology* (2021; a translated and updated version of a title originally published in Norwegian). This book draws closely on the authors' work in Africa and Norway, and it offers great further reading on development-related themes.

In 2015, two political ecology "handbooks" were published. They are particularly useful for postgraduate and postdoctoral researchers, offering extensive bibliographies and inspiration for further research. Raymond Bryant's *International Handbook of Political Ecology* (2015b) and Tom Perreault et al.'s *Routledge Handbook of Political Ecology* (2015) both comprise just short of 50 chapters, each of which takes on a select issue and reviews the literature in-depth. Rebecca Lave et al.'s *Palgrave Handbook of Critical Physical Geography* (2018) is a brilliant addition to an allied field.

There are currently five "in-house" journals in political ecology. *Journal of Political Ecology* is a central publication venue (and was also one of the world's first open-access journals!). *Environment and Planning E: Nature and Space* and the "Environment, Politics and Society" section of *Frontiers in Human Dynamics* were established more recently. *Écologie et Politique* publishes in French and *Ecología Política* in Spanish. In addition to these journals, *Progress in Human Geography* issues a political ecology "Progress report" regularly. The reports are outstanding reading if you want to keep track of the most recent developments in the field. If you read through the bibliographies at the end of each chapter in this book, you will soon also find patterns indicating which anthropology, geography and other general-field journals are worth exploring.

BIBLIOGRAPHY

Adams, W. M. 2020. *Green Development: Environment and Sustainability in a Developing World*, 4th ed. London, UK: Routledge.

Agrawal, Arun. 2005. *Environmentality: Technologies of Government and the Making of Subjects*. Durham, NC: Duke University Press.

Alexievich, Svetlana. 2016 (1997). *Chernobyl Prayer: A Chronicle of the Future*. Translated by Anna Gunin and Arch Tait. London, UK: Penguin.

Alimonda, Héctor, Catalina Toro Pérez, and Facundo Martín, eds. 2017. *Ecología política: Pensamiento crítico, diferencia latinoamericana y rearticulación epistémica* [*Political Ecology: Critical Thought, Latin American Difference, and Epistemic Rearticulation*]. Buenos Aires, Argentina: CLASCO.

Arora-Jonsson, Seema, Carol J. Pierce Colfer, and Marien González-Hidalgo. 2021. "Seeing the Quiet Politics in Unquiet Woods: A Different Vantage Point for a Future Forest Agenda." *Human Ecology* 49: 297–308.

Baker, Kate, Markus P. Eichhorn, and Mark Griffiths. 2019. "Decolonizing Field Ecology." *Biotropica* 51(3): 288–292.

Bakker, Karen. 2012. "Water: Political, Biopolitical, Material." *Social Studies of Science* 42(4): 616–623.

Bakker, Karen, and Gavin Bridge. 2021. "Material Worlds Redux: Mobilizing Materiality within Critical Resource Geography." In *Routledge Handbook of Critical Resource Geography*, edited by Matthew Himley, Elizabeth Havice, and Gabriela Valdivia, 43–56. London, UK: Routledge.

Barry, Andrew. 2013. *Material Politics: Disputes along the Pipeline*. Malden, MA: Wiley-Blackwell.

Batterbury, Simon. 2015. "Doing Political Ecology inside and outside the Academy." In *International Handbook of Political Ecology*, edited by Raymond L. Bryant, 27–43. Cheltenham, UK: Edward Elgar.

Benjaminsen, Tor A., and Hanne Svarstad. 2017. *Politisk økologi: Miljø, mennesker og makt*, 2nd ed. [*Political Ecology: Environment, People and Power*]. Oslo, Norway: Universitetsforlaget.

Benjaminsen, Tor A., and Hanne Svarstad. 2021. *Political Ecology: A Critical Engagement with Global Environmental Issues*. Cham, Switzerland: Palgrave Macmillan.

Blaikie, Piers. 1985. *The Political Economy of Soil Erosion in Developing Countries*. London, UK: Longman.

Blaikie, Piers, and Harold Brookfield. 1987. "Defining and Debating the Problem." In *Land Degradation and Society*, edited by Piers Blaikie and Harold Brookfield, 1–26. London, UK: Methuen.

Blaser, Mario. 2014. "Ontology and Indigeneity: On the Political Ontology of Heterogenous Assemblages." *cultural geographies* 21(1): 49–58.

Braun, Bruce. 2006. "Towards a New Earth and a New Humanity: Nature, Ontology, Politics." In *David Harvey: A Critical Reader*, edited by Noel Castree and Derek Gregory, 191–222. Malden, MA: Blackwell.

Bryant, Raymond L. 2015a. "Reflecting on Political Ecology." In *International Handbook of Political Ecology*, edited by Raymond L. Bryant, 14–24. Cheltenham, UK: Edward Elgar.

Bryant, Raymond L., ed. 2015b. *International Handbook of Political Ecology*. Cheltenham, UK: Edward Elgar.

Bryant, Raymond L., and Sinéad Bailey. 1997. *Third World Political Ecology*. London, UK: Routledge.

Cederlöf, Gustav. 2016. "Low-Carbon Food Supply: The Ecological Geography of Cuban Urban Agriculture and Agroecological Theory." *Agriculture and Human Values* 33(4): 771–784.

Cederlöf, Gustav. 2021. "Out of Steam: Energy, Materiality, and Political Ecology." *Progress in Human Geography* 45(1): 70–87.

Chartier, Denis, and Estienne Rodary. 2015. "Globalizing French *Écologie Politique*: A Political Necessity." In *International Handbook of Political Ecology*, edited by Raymond L. Bryant, 547–560. Cheltenham, UK: Edward Elgar.

Daigle, Michelle, and Juanita Sundberg. 2017. "From Where We Stand: Unsettling Geographical Knowledges in the Classroom." *Transactions of the Institute of British Geographers* 42(3): 338–341.

D'Alisa, Giacomo, Federico Demaria, and Giorgos Kallis, eds. 2014. *Degrowth: A Vocabulary for a New Era*. London, UK: Routledge.

Davis, Janae, Alex A. Moulton, Levi Van Sant, and Brian Williams. 2019. "Anthropocene, Capitalocene, … Plantationocene? A Manifesto for Ecological Justice in an Age of Global Crises." *Geography Compass* 13: e12438.

Durand, Leticia, and Juanita Sundberg. 2022. "Monster Plants: The Vegetal Political Ecology of *Lacandonia schismatica*." *Journal of Political Ecology* 29: 189–207.

Escobar, Arturo. 1995. *Encountering Development: The Making and Unmaking of the Third World*. Princeton, NJ: Princeton University Press.

Escobar, Arturo. 1999. "After Nature: Steps to an Antiessentialist Political Ecology." *Current Anthropology* 40(1): 1–30.

Freitas, Tiago Ávila Martins, and Augusto Cesar Salomão Mozine. 2015. "Towards a Lusophone Political Ecology: Assessing '*Para inglês ver*' Environments." In *International Handbook of Political Ecology*, edited by Raymond L. Bryant, 604–618. Cheltenham, UK: Edward Elgar.

Gandy, Matthew. 2022. *Natura Urbana: Ecological Constellations in Urban Space*. Cambridge, MA: MIT Press.

Gautier, Denis, and Christian A. Kull. 2015. "French Research Traditions of Peasant Agricultural Systems: A Convergence with Political Ecology?" In *Routledge Handbook of Political Ecology*, edited by Tom Perreault, Gavin Bridge, and James McCarthy, 76–88. London, UK: Routledge.

Givens, Jennifer E., Xiaorui Huang, and Andrew K. Jorgensen. 2019. "Ecologically Unequal Exchange: A Theory of Global Environmental *In*justice." *Sociology Compass* 13(5): e12693.

Grosfoguel, Ramón. 2013. "The Structure of Knowledge in Westernized Universities: Epistemic Racism/Sexism and the Four Genocides/Epistemicides of the Long 16th Century." *Human Architecture: Journal of the Sociology of Self-Knowledge* 11(1): 73–90.

Gudynas, Eduardo. 2016. "Beyond Varieties of Development: Disputes and Alternatives." *Third World Quarterly* 37(4): 721–732.

Harcourt, Wendy, and Ingrid L. Nelson, eds. 2015. *Practising Feminist Political Ecologies: Moving beyond the "Green Economy"*. London, UK: Zed.

Heynen, Nik, Maria Kaika, and Erik Swyngedouw, eds. 2006. *In the Nature of Cities: Urban Political Ecology and the Politics of Urban Metabolism*. London, UK: Routledge.

Hjorth Warlenius, Rikard. 2022. "Learning for Life: ESD, Ecopedagogy and the New Spirit of Capitalism." *Journal of Environmental Education* 53(3): 141–153.

Jönsson, Erik, and Elina Andersson, eds. 2017. *Politisk ekologi: Om makt och miljöer* [*Political Ecology: On Power and Environments*]. Lund, Sweden: Studentlitteratur.

Kahn, Richard. 2010. *Critical Pedagogy, Ecoliteracy, and Planetary Crisis: The Ecopedagogy Movement*. New York, NY: Peter Lang.

Kallis, Giorgos, Susan Paulson, Giacomo D'Alisa, and Federico Demaria. 2020. *The Case for Degrowth*. Cambridge, UK: Polity Press.

Karlsson, Bengt G. 2015. "Political Ecology: Anthropological Perspectives." In *International Encyclopedia of the Social & Behavioral Sciences* 18, 2nd ed., edited by James D. Wright, 350–355. Oxford, UK: Elsevier.

Kim, Soyeun, Godwin Uyi Ojo, Rukhe Zehra Zaidi, and Raymond L. Bryant. 2012. "Bringing the Other into Political Ecology: Reflecting on Preoccupations in a Research Field." *Singapore Journal of Tropical Geography* 33: 34–48.

Kothari, Ashish, Ariel Salleh, Arturo Escobar, Federico Demaria, and Alberto Acosta, eds. 2019. *Pluriverse: A Post-Development Dictionary*. New Delhi, India: Tulika Books.

Lave, Rebecca. 2015. "Reassembling the Structural: Political Ecology and Actor-Network Theory." In *Routledge Handbook of Political Ecology*, edited by Tom Perreault, Gavin Bridge, and James McCarthy, 213–223. London, UK: Routledge.

Lave, Rebecca, Christine Biermann, and Stuart N. Lane, eds. 2018. *The Palgrave Handbook of Critical Physical Geography*. Cham, Switzerland: Palgrave Macmillan.

Leff, Enrique. 2015. "Political Ecology: A Latin American Perspective." *Desenvolvimento e Meio Ambiente* 35: 29–64.

Li, Tania Murray. 2007. *The Will to Improve: Governmentality, Development, and the Practice of Politics*. Durham, NC: Duke University Press.

Loftus, Alex. 2019. "Political Ecology 1: Where Is Political Ecology?" *Progress in Human Geography* 43(1): 172–182.

Martinez-Alier, Joan. 2002. *The Environmentalism of the Poor: A Study of Ecological Conflicts and Valuation*. Cheltenham, UK: Edward Elgar.

Metcalf, Thomas R. 1995. *Ideologies of the Raj*. Cambridge, UK: Cambridge University Press.

Nagendra, Harini. 2016. *Nature in the City: Bengaluru in the Past, Present, and Future*. New Delhi, India: Oxford University Press.

Newell, Joshua P., Joshua J. Cousins, and Jennifer Baka. 2017. "Political-Industrial Ecology: An Introduction." *Geoforum* 85: 319–323.

Öhman, May-Britt. 2016. "TechnoVisions of a Sámi Cyborg: Reclaiming Sámi Body-, Land-, and Waterscapes after a Century of Colonial Exploitations in Sábme." In *Illdisciplined Gender: Engaging Questions of Nature/Culture and Transgressive Encounters*, edited by Jacob Bull and Margaretha Fahlgren, 63–98. Cham, Switzerland: Springer.

Otero, Iago, Giorgos Kallis, Raül Aguilar, and Vicenç Ruiz. 2011. "Water Scarcity, Social Power and the Production of an Elite Suburb: The Political Ecology of Water in Matadepera, Catalonia." *Ecological Economics* 70(7): 1297–1308.

Paulson, Susan, Lisa L. Gezon, and Michael Watts. 2003. "Locating the Political in Political Ecology: An Introduction." *Human Organization* 62(3): 205–217.

Peluso, Nancy Lee. 1992. "The Political Ecology of Extraction and Extractive Reserves in East Kalimantan, Indonesia." *Development & Change* 23(4): 49–74.

Perreault, Tom, Gavin Bridge, and James McCarthy, eds. 2015. *Routledge Handbook of Political Ecology*. London, UK: Routledge.

Quijano, Aníbal. 2000. "Coloniality of Power, Eurocentrism, and Latin America." *Nepantla: Views from South* 1(3): 533–580.

Robbins, Paul. 2020. *Political Ecology: A Critical Introduction*, 3rd ed. Oxford, UK: Wiley-Blackwell.

Rocheleau, Diane, Barbara Thomas-Slayter, and Esther Wangari, eds. 1996. *Feminist Political Ecology*. London, UK: Routledge.

Said, Edward W. 1979. *Orientalism: Western Conceptions of the Orient*. New York, NY: Vintage Books.

Schmid, Sonja D. 2015. *Producing Power: The Pre-Chernobyl History of the Soviet Nuclear Industry*. Cambridge, MA: MIT Press.

Spash, Clive L., ed. 2017. *Routledge Handbook of Ecological Economics: Nature and Society*. London, UK: Routledge.

Sultana, Farhana. 2021. "Political Ecology 1: From Margins to Center." *Progress in Human Geography* 45(1): 156–165.

Truelove, Yaffa. 2019. "Gray Zones: The Everyday Practices and Governance of Water beyond the Network." *Annals of the American Association of Geographers* 109(6): 1758–1774.

Tsing, Anna L. 2015. *The Mushroom at the End of the World: On the Possibility of Life in Capitalist Ruins*. Princeton, NJ: Princeton University Press.

Ulloa, Astrid. 2015. "Environment and Development: Reflections from Latin America." In *Routledge Handbook of Political Ecology*, edited by Tom Perreault, Gavin Bridge, and James McCarthy, 320–331. London, UK: Routledge.

Valdivia, Gabriela. 2015. "Eco-Governmentality." In *Routledge Handbook of Political Ecology*, edited by Tom Perreault, Gavin Bridge, and James McCarthy, 467–480. London, UK: Routledge.

Vayda, Andrew P. 1983. "Progressive Contextualization: Methods for Research in Human Ecology." *Human Ecology* 11(3): 265–281.

Vayda, Andrew P., and Bradley B. Walters. 1999. "Against Political Ecology." *Human Ecology* 27(1): 167–179.

Walker, Peter A. 2005. "Political Ecology: Where Is the Ecology?" *Progress in Human Geography* 29(1): 73–82.

Watts, Michael J. 2013 (1983). *Silent Violence: Food, Famine, and Peasantry in Northern Nigeria*. Athens, GA: University of Georgia Press.

Watts, Michael, and Richard Peet. 2004. "Liberating Political Ecology." In *Liberation Ecologies*, 2nd ed., edited by Richard Peet and Michael Watts, 3–43. London, UK: Routledge.

Watts, Vanessa. 2013. "Indigenous Place-Thought & Agency amongst Humans and Non-Humans (First Woman and Sky Woman Go on a European World Tour!)." *Decolonization: Indigeneity, Education & Society* 2(1): 20–34.

Yeh, Emily T. 2015. "Political Ecology in and of China." In *International Handbook of Political Ecology*, edited by Raymond L. Bryant, 619–632. Cheltenham, UK: Edward Elgar.

CHAPTER 2
Power

When the study of ecology first began in the late nineteenth century, it challenged what long had been a dominant Western understanding of the natural world. The biologists were examining the internal coherence of cells and individual organisms, but the new science rested on the assumption that an organism could not be fully understood on its own—apart from its interactions with other organisms and the environment. This was a radical shift in perspective. An individual could only meaningfully be studied in relation to others. An organism, a body or even a machine had to be seen as an ensemble of social and material relations that extended beyond it in space and time.

The ecological perspective had consequences also in the social sciences and the humanities. Scholars in the field of cultural ecology, trained in anthropology and geography, started to examine how livelihoods and forms of social organisation resulted from the cultural and physical interactions between humans and their environment (Rappaport 1968; Nietschmann 1973). But with humans inserted into the ecosystems under study, it soon became evident that the differences and unequal relations that existed within and between social groups were a key element in human-environment interactions. Who owned the natural resources in an area, and who had the ability to benefit from them? Who controlled the material flows out of which the environments for human action were constituted, and what social and cultural mechanisms sustained those flows? Whose knowledge of the environment counted, and whose way of life was prioritised in the face of environmental crisis? The unequal relations that shaped human interactions with the environment were questions of power.

For political ecologists, power is a keyword. The "political" in political ecology, as Susan Paulson et al. (2005: 28) write, refers not only to politics as it takes place in parliaments and international forums but rather designates "the practices and processes through which power, in its multiple forms, is wielded and negotiated". More often than not, the concept is implicit in conversations in the field, but in this chapter, we identify five main ways in which the idea is understood. The different notions have their distinct intellectual lineages, emerging from different schools of thought, but rather than treat them as contradictory, we see them as complementary. One concept gives us a particular perspective on human-environment interactions, which highlights certain patterns while occluding others. One notion will be more useful than another depending on the nature of the questions we ask. And yet, as Hanne Svarstad et al. (2018: 350) argue, the ability to employ multiple perspectives on power simultaneously is a key

DOI: 10.4324/9781003095521-2

strength in political ecology and provides for a nuanced understanding of how unequal social relations are manifest in the environment.

POWER AS A RESOURCE

A basic way of conceptualising power is to view it as an attribute of a person or an organisation. From this perspective, the environment can be seen as a stage on which actors of different strengths enter into conflict. The political scientist Robert Dahl (1957: 202–203) gave a classic definition of such an actor-centred perspective, stating concisely that "A has power over B to the extent that A can get B to do something that B would not otherwise do". Many texts on power begin with a reference to Dahl, just like this, and also the work of Steven Lukes (2005)—a political theorist whose work is closely associated with this understanding of power. Following the established canon, it is difficult to ground a discussion of the different ways in which power is understood without reference to a set of male European scholars such as Dahl and Lukes. However, while recognising the importance of their contributions, it is also our ambition to bring attention to how power is understood in far more diverse contexts and by a far more diverse array of people than this.

When a group of women belonging to the Standing Rock Sioux Tribe set up camp to protest the construction of the Dakota Access Pipeline in April 2016, they argued that the pipeline would desecrate their ancestral burial grounds. If the pipeline would leak (which is not uncommon for pipelines to do), it would also imperil the water supply to their community. On social media, the hashtags #NoDAPL, #Mniwiconi and #Waterislife started trending, and the resistance movement grew fast, becoming the largest assembly of native people in the history of the United States. On 3 September 2016, construction workers arrived on site with excavators, and the nonviolent activists clashed with security forces who were equipped with mace and attack dogs over the right to land (Estes 2019; Gilio-Whitaker 2019). It is easy to picture power as a zero-sum game at Standing Rock in which different environmental actors held a greater or lesser degree of power, which they exercised as a tactical resource.

In *Third World Political Ecology*, Raymond Bryant and Sinéad Bailey (1997) developed an understanding of environmental conflict based on this perspective of power and outlined multiple ways in which actors exert control over the environments of others. Two ways stand out in particular. First, a powerful actor can seek to regulate the access to land, resources and bodies in an area, for example with the objective of piping oil between sites of production and consumption or to monopolise the commercial value of a resource. Bryant and Bailey identify colonial forestry as a case in point of the latter. Until the late nineteenth century, local communities in South Asia's forests sustained their livelihoods through hunting, gathering and pastoralism, but as we discuss at greater length in Chapter 8, European mercantile corporations and later colonial states placed large forest tracts under their control. By codifying property laws and declaring forests state property, local communities were deemed trespassers on state land and their customary practices were restricted to the benefit of the timber industry (Figure 2.1; Rangarajan and Sivaramakrishnan 2012).

Second, an actor may seek to prioritise and enforce its own environmental project on that of another. For instance, the construction of photovoltaic solar parks has recently become a contentious issue when states and businesses dispossess rural communities of land and assets for low-carbon development (Yenneti et al. 2016). The idea of carbon trading (Chapter 8) similarly

FIGURE 2.1 A sign demarcating state property in Uttarakhand, India

While walking in a forest in northern India, one of us came across this sign, which manifests the power of the state's Forest Department in the landscape. In the same way, walls, hedgerows, buildings, dams or other infrastructures can make one actor's power over another spatially manifest (Duncan 1989).

Photo by Gustav Cederlöf

enables resource-strong actors, often in the global north, to displace the environmental impact of their consumption to the global south, where carbon-capture forestry projects marginalise small farmers and pastoralists who depend on forest access to sustain their livelihoods (Asiyanbi 2016).

In modern societies, "access" is conventionally defined as "the *right* to benefit from things" (Ribot and Peluso 2003). The notion of a right, in turn, is closely connected with the state as an actor, as rights only exist in the context of a society in which they are enforceable through law, custom or convention. People can access resources legally by asserting property ownership or by approaching rights holders to gain access, but also illegally through theft or pilfering. Corruption is another widespread phenomenon regulating access to land and resources, which can be seen as an informal, extra-legal institution rooted in local systems of power (Robbins 2000).

Theorising access, however, Jesse Ribot and Nancy Lee Peluso (2003) argue that a narrow focus on rights obscures the complex ways in which people actually come to benefit from natural resources. They instead define access as "the *ability* to benefit from things", which turns access into a question of power. Power, they argue, operates in three domains to regulate access: first, in terms of control, which is an actor's ability to mediate someone else's access to nature. Second, in terms of maintenance, which draws attention to the economic, cultural, legal and infrastructural work necessary to sustain a particular access arrangement. And third, in terms of attainment, which is the process by which people gain access through those who have control.

When power is seen as a resource, it is easy to view it as one's "power over" another. Nevertheless, many political ecologists have studied and foregrounded the mirror image of domination: resistance. While environmental social movements, such as the one at Standing Rock, represent overt protest to a more powerful actor, early research drew inspiration from studies of less visible forms of resistance. In *Weapons of the Weak*, James Scott (1985) recorded how peasants, whose life situation often makes them risk averse, tended to resist the domination of landowners and state institutions not by staging rebellion but through a prosaic, continuous struggle. "[I]t is rare for peasants to risk an outright confrontation with the authorities over taxes, cropping patterns, development policies, or onerous new laws", Scott argued and continued:

> instead they are likely to nibble away at such policies by noncompliance, foot dragging, deception. In place of land invasion, they prefer piecemeal squatting; in place of open mutiny, they prefer desertion; in place of attacks on public or private grain stores, they prefer pilfering.
>
> (xvi)

Political ecologists have subsequently demonstrated how both rural and urban dwellers resist domination in a range of mundane activities, for example by poaching wildlife in national parks for subsistence or commercial gain (Duffy et al. 2016) or by reengineering water and electricity meters to bypass them and contest the neoliberal logic they enact (von Schnitzler 2013). By viewing political struggle from the perspective of the resource-weak, a positive way of thinking about power, instead of someone's "power over", is in terms of a person's "power to" act and resist; their "power with" others to effect change; and their "power within", referring to an individual's self-esteem, awareness and resolve (VeneKlasen and Miller 2002).

POWER AS A SOCIAL RELATION

The resource perspective on power implies that someone holds power and that this someone exists as an autonomous actor in a given environment. However, as Ribot and Peluso (2003) point out, an actor's access to resources is always mediated by less tangible cultural and political economic mechanisms that extend beyond any one actor. A second way of thinking about power draws attention away from the individual as an actor intentionally wielding power and instead focuses on the social relations that enable and constrain human agency (Table 2.1). When Anglo-American political ecology first developed in the 1980s and 1990s, Marxist political economy was a particularly productive perspective, offering conceptual tools to examine how power relations shape the circumstances in which individuals behave as economic actors. As we noted in Chapter 1, influential studies of famine, soil erosion and deforestation demonstrated how conflicts over access to natural resources were rooted in systems of political and economic

TABLE 2.1 Five definitions of power

Power as...	Focus	Tradition(s)	Associated concepts
Resource	Actors	Various	Power over, power to, power with, power within
Social relation	Structures	Marxism, feminism, intersectionality	Class, kinship, gender, race, ethnicity, caste, age, ability, migrant status, etc.
Ideology	Maintenance of the status quo	Marxism (Gramsci)	Common sense, hegemony, praxis
Discourse	Meaning, subjectivity	(Post-)structuralism	Sign, signification, social construction, discursive struggle
Knowledge	Classification	Foucauldian	power/knowledge, taxonomy, coloniality of knowledge, epistemologies of the south

control that needed to be understood in a long-term perspective, often with colonial roots (Blaikie 1985; Schmink and Wood 1992; Watts 2013). These studies also showed how a person's ability to act was contingent on their position in a wider set of social circumstances; circumstances over which they were not fully in control. To conceptualise power as a social relation, these circumstances are where we need to look.

Karl Marx may not be widely known as an environmental theorist, but his work is firmly grounded in two assumptions about how humans relate to the environment. First, Marx notes that humans have an active relation to nature, transforming it through labour to sustain their livelihoods. Labour, he writes, "is a condition of human existence which is independent of all forms of society; it is an eternal natural necessity which mediates the metabolism between man [*sic*] and nature, and therefore human life itself" (Marx 1976: 133). Indeed, to transform nature, humans employ tools, machines, infrastructures and other species—what Marx calls the "means of production"—to surpass our organic limitations.

Second, humans live as part of larger social formations, and Marx argues that the patterns of socio-economic organisation—the "relations of production"—shape how we transform nature through labour. To examine the relations of production at a specific point in space and time, we can ask a set of *who*-questions about the distribution of resources, land, technology and produce:

1 Who owns the means of production?
2 Who operates them?
3 Who controls produce that materially reproduces the social formation?
4 Who controls surplus produce and under what conditions?

In Marx's analysis of capitalist production in the Victorian era, he demonstrated how the producers tended to be separated from the means of production and agreed to operate them in return for a wage. Yet this was not entirely by their own volition. Lacking means of production, the producers were under economic pressure to sell their labour in return for a wage with which they could purchase commodities for consumption. By contrast, the owners of the means of production needed to purchase labour in order to produce commodities, and they were able to do so by controlling surplus produce and selling it for more money than they paid in return for labour. Marx's conclusion was that two social classes came to dominate a capitalist economy: owners and producers—capitalists and wage labourers. The relations of production, in turn, exerted power on members of these classes, compelling them to act in certain ways as they were situated in a socio-economic relationship to others. Meantime, the relations of production specified a form of social organisation that governed human-environment interactions.

In the 1970s and 1980s, Marxist historians engaged in heated discussions as they tried to understand how European capitalism had developed out of qualitatively different agrarian class structures (Aston and Philpin 1985). In an agrarian subsistence economy, the owners of the means of production also tend to operate them and use what they produce, including any surpluses, to sustain their livelihoods. Under relations of servitude in the feudal economies of medieval Europe, they were also forced to transfer surplus produce to the nobility in return for protection and land for subsistence farming. In comparison, more recent studies in political ecology have examined how the so-called Green Revolution of agricultural mechanisation and intensification, which dominated development policy in the period after the Second World War, has incorporated peasant farmers in relations of market exchange (Shiva 2014; Dove and Kammen 2015). Another important area of inquiry has been how the currently ongoing commodification of carbon emissions and ecosystem services (Chapter 4) transform social relations by inserting people, places and resources into patterns of neoliberal capitalist production (Peluso 2012; Fletcher et al. 2015).

Marxist perspectives remain influential in the field, but political ecologists—and often those who are women and non-Western scholars—have long stressed that readings of power as an unequal social relation defined by class can result in mechanistic interpretations of society. As an immanent critique, two influential Marxist political ecologists also argued early on that there is a risk with Marxist analyses that people, with our ideas, emotions and actions, are reduced to "passive 'bearers' of social relations" (Peet and Watts 1996: 28). The challenge, as the historian E. P. Thompson (1980: 8) famously argued, is to study class as a lived historical phenomenon "which in fact happens (and can be shown to have happened) in human relationships". A historically open perspective on class such as this is necessarily also attentive to other power relations that govern human-environment interactions.

Building on this critique, political ecology is often characterised by a broader "intersectional" understanding of power as a social relation today (Crenshaw 1991). For example, beyond class, kinship is one prominent hierarchical system that regulates access to resources based on notions of filiation and affinity, placing people in social relations that permit them to call upon the labour and produce of others (Wolf 1982). Gender is another axis of difference that shapes people's control over resources, and feminists have been particularly influential in theorising power in political ecology (Rocheleau et al. 1996; Harcourt and Nelson 2015; Sundberg 2017). While initially tending to focus on gender alone, feminist political ecologists now stress that power relations of different kinds cannot be isolated in a complex world (Sultana 2021). Class, kinship and gender are not three social relations that exist independently of each other, but rather, different social identities intersect to shape people's material, symbolic and emotional engagements with nature.

In New Delhi, about half the population lacks a formal connection to the city's centralised water network. In a recent study, Yaffa Truelove (2019: 1759) explains how people instead install illegal pumps and pipes, purchase water privately or ask neighbours for an extra bucket. They thus go "beyond the network" to make water flow across the city. In an informal South Delhi settlement, Truelove examines how an NGO helped to reactivate an old tube well. The mundane household task of collecting water usually fell to women in the area, she explains, and a women's collective formed to manage the new infrastructure. However, while inclusive of some women, the NGO excluded residents from the mostly Muslim side of the settlement in the management arrangement, and this differentiated the experience of water access not only in gendered but also in ethno-religious terms.

The area's municipal representative later agreed to sponsor the tube well on condition that the local chieftain (the *Pradhan*) managed the infrastructure. Yet the *Pradhan* allotted more water to dominant-caste Brahmin families from whom he also had political support while restricting access to families of lower castes. Muslim men were the most disadvantaged, Truelove concludes, who tended to live alone in the area, having migrated to the city for work. While excluded on ethno-religious terms, they also had to shoulder the typically feminised task of water collection even as they were absent from the area at daytime to wage labour. Relations of gender, ethno-religion, class, caste and patronage thereby wove into an intricate social fabric shaping access and the experience of accessing water.

The intersectional perspective on power assumes that identities do not exist outside social praxis, as pre-given categories, but that they are performed and hence emerge from historically and culturally situated interactions (Butler 1990). Andrea Nightingale (2011: 153) argues that the performance of social identities in human-environment interactions, such as in the governance of water infrastructure, "collapse[s] the distinction between the material and the symbolic". Power, therefore, ties together the social and the natural, working across any such analytical distinction, and is located in the embodied work practices that produce identities, bodies and environments. As a social relation providing the circumstances for human action, an intersectional view on power can be summarised in four points:

1. It is *relational*, enabling and constraining human action by creating social difference.
2. It is *coincidental* in that the multiple forms of social difference through which power operates exist at the same time and intersect with each other. These forms include but are not limited to class, kinship, gender, sexuality, ethnicity, religion, race, caste, patronage, immigration status, age, health, ability, education and profession.
3. It is *contingent*, implying that human relationships are subject to change and in the process of becoming.
4. It is *embodied*, suggesting that power relations are rooted in everyday practices in which they are repeatedly lived out.

Despite these insights, Sharlene Mollett and Caroline Faria (2013) remark that race has received relatively little attention from feminist political ecologists, perhaps reflecting the relative whiteness of the Anglo-American research community. When discussing race, it is important to distinguish between "race" as a social system of inequality that has developed historically with roots in European colonialism and "races" as actually existing, observable categories of people. While the former is a historical phenomenon with political implications, which we can study empirically in diverse contexts, the latter is an obsolete and untenable way of representing social difference. A useful distinction is instead between "race" as a contingent historical phenomenon and "racialisation" as the process in which racial differences are symbolically and materially produced at different times in different places (Appelbaum et al. 2003: 3).

Latin American political ecologists have been extra prominent in theorising race. Grounded in their experiences as Latin Americans, their efforts have also been connected with a radical critique of the global economy. Notably, the Peruvian sociologist Aníbal Quijano (2000) argues that it is not enough to work with a concept of class to understand the world economy—we also need a concept of race. The idea of race originated with the Spanish and Portuguese conquest of the Americas when the continent, through looting, land occupation and resource extraction, was rendered an inferior place to Europe. The Americas was a place in which racialised bodies justifiably could be exploited and from which natural resources legitimately could be extracted (Figure 2.2). As an existing system of inequality, Quijano argues that race divides

FIGURE 2.2 The Cerro Rico in Potosí, Bolivia

In *Open Veins of Latin America*, a seminal book for the early development of Latin American political ecology, Eduardo Galeano (1997: 40–41) writes about the Spanish regime of silver mining in Potosí:

"The *mita* labor system was a machine for crushing Indians. The process of using mercury to extract silver poisoned as many or more than did the toxic gases in the bowels of the earth. It made hair and teeth fall out and brought on uncontrollable trembling. The victims ended up dragging themselves through the streets pleading for alms. At night 6,000 fires burned on the slopes of the Cerro and in these the silver was worked, taking advantage of the wind that the 'glorious Saint Augustine' sent from the sky. Because of the smoke from the ovens there were no pastures or crops for a radius of twenty miles around Potosí and the fumes attacked men's bodies no less relentlessly".

Photo by iStock.com/rchphoto

the world population based on Eurocentric rationality, which assumes that all humans share a common history of modernisation. This universal history has culminated in the technological and economic development of Euro-America, and within it, differences between Europe and non-Europe are explained as the outcome of natural (racial) differences and in terms of non-European countries "lagging behind" and "lacking development" (Escobar 1995; Leff 2015). To the contrary, Quijano contends that differences in space that exist *at the same time* cannot be explained as differences *in time* but must be explained in relation to each other as coincident phenomena. Western modernity must be understood in relation to colonial forms of domination which persist today despite the end of formal colonialism. If modernity is one side of the coin, he argues, "coloniality" (*colonialidad* in Spanish) is the other. And yet, to bring the argument full circle, coloniality intersects with other systems of power to position people unevenly in global relations of production (Grosfoguel 2007).

POWER AS IDEOLOGY

The discussion so far illustrates how power operates through the circumstances in which people act. "Rather than A getting B to do something which B would not otherwise do", Jeffrey Isaac (1987: 25) writes to echo Dahl's classic definition, "social relations of power typically involve both A and B doing what they ordinarily do". In these terms, a lack of resistance does not indicate a lack of power but power's very success. An important question, then, is why unequal power relations persist if they are unjust. How and why are power relations maintained in everyday life? While the resource perspective on power draws attention to different forms of coercion that we discussed earlier, a less salient answer builds on a third way in which political ecologists conceptualise power: as ideology.

Ideology can be conceptualised in several ways. Reflecting on these different interpretations, the cultural critic Raymond Williams (1983: 157, original emphasis) suggests the most common interpretation is a negative one: "Sensible people", he writes, "rely on experience, or have a *philosophy*, silly people rely on ideology". This negative framing is notably evident in Henri Lefebvre's (1982: 64) work when he writes that ideologies are "an inverted, truncated, distorted reflection of reality". Political ecologists have tended to work with ideology in this sense as a way of understanding how particular unequal socio-ecological formations are rendered natural or normalised. The power of ideology is to make things appear impossible to change. For example, Juanita Sundberg (2008: 570 citing Goldberg 1993, emphasis added) argues that the power of race rests in how it "both establishes and rationalizes the order of difference as *a law of nature*". Mukul Sharma (2018) instead shows how caste is naturalised and negotiated in India when, for example, water is understood to be "polluted" as soon as a Dalit touches it. In his critique of Thomas Malthus, David Harvey (1974) similarly sees Malthusian interpretations of the relationship between population growth and resource scarcity as a kind of ideological false thought (Chapter 4).

Paul Robbins (2020) argues that narratives that explain social differences and environmental problems with reference to "natural causes" are among the most powerful forces to sustain the status quo. One widespread narrative that confers explanatory power to "natural causes" turns on the notion that disasters are caused by nature "out there". This narrative would suggest that what makes you vulnerable to a hazard, such as an earthquake, wildfire or typhoon, is that you live in a "dangerous" environment. However, if we analyse disasters through a power lens, it is evident that a person's or a community's ability to prepare for, cope with, respond to and recover from a disaster is a question of social vulnerability determined by relations of class, gender, age, mobility, etc. (Wisner et al. 2004). And yet, a persistent ideology of nature "out there" eclipses such analyses and sustains the political economic processes that generate vulnerability by concealing them (Chapter 6).

Somewhat confusingly, ideology can also be understood in two other main senses: we might speak about ideology in positive terms, for example seeing environmentalism as an ideology concerned with improving the environment. And we might regard it in a more neutral sense, as a terrain upon which different interpretations of the world are struggled over. Political ecologists trained in anthropology often emphasise the central role of culture in human-environment interactions, and culture can be understood as a socially negotiated system of meaning through which humans classify and make sense of the world. Several political ecologists have drawn on the Italian Marxist Antonio Gramsci in exploring ideology as meaning-making (Moore 1996; Ekers et al. 2013; Karriem 2013). Gramsci used the term ideology in all three of the senses discussed above (and more still), and his main interest was to understand how conceptions of the world can be translated into a coherent philosophy and praxis with the potential of bringing about radical change.

Gramsci lived through a period of intense revolutionary optimism in Italy, which gave way to Mussolini's fascist political project. He was, therefore, deeply concerned with this project of "translation". A scholar of linguistics, Gramsci's understanding of translation was related to the use of language (Ives 2004), but he also saw it as a collective project that involved working with and learning from the fragmented understandings that, in a particular context, constituted "common sense", transforming these into "good sense" and praxis (Hart 2013). Thus, Gramsci's *Prison Notebooks* came to inspire what was known as the Subaltern Studies Collective—a group of scholars in India who attempted to write history "from below" and understand the world through the common sense of marginalised groups. Subaltern studies have later exerted a considerable, if often unacknowledged influence on political ecology, and it is perhaps only a slight exaggeration when one of us once claimed that Gramsci may well be considered "the pre-eminent theorist of 'political ecology' within the Marxist tradition" (Loftus 2013: 194).

POWER AS DISCOURSE

A closely related fourth way in which political ecologists conceptualise power resembles the Gramscian perspective but emerged from one of the other great intellectual movements of the twentieth century: structuralism. Structuralist thinking starts with the observation that no natural connection exists between the sound or written image of a word, such as "tree", and the idea that we picture when we hear the word—i.e. the plant with roots, trunk and foliage. The Swiss linguist Ferdinand de Saussure argued that it is only due to social convention that we learn about such a connection. Saussure's idea was that a sign, such as "tree", only attains meaning by being part of a larger structure—a language—in which it is different from other signs. We know what a "tree" is because we know what a "root" is and because we know what "trunk" and "foliage" are. Meaning, therefore, is not immanent but contextual. As Ania Loomba (2015: 53) argues, "language is not a nomenclature, or a way of naming things which already exist, but a system of signs, whose meaning is relational". Today, we refer to such a system of signs as a discourse.

The Van Gujjar is a group of nomadic pastoralists who live as buffalo herders in the Central Himalayas. They sell milk to nearby settled communities for income, and on a seasonal basis, they migrate from the forests in the plains (known as *khols*) to pastures at altitudes above 3,000 metres. For the Van Gujjar, the forest is the wealth of a home. It is a place of life and death and the source of identity and memory. The paths along the riverbeds (known as *raos*) connect families with friends and lead to the market and the hills. Pernille Gooch (2004: 129, original emphasis) explains how the Van Gujjar understand the region they inhabit as an interconnected whole:

> [I]n summer the force of life itself moves from the dried out forest in the foothills up into the new fresh pastures in the mountains. After summer, life moves the other way back into the lowland where everything will be fresh and new after the monsoon, while the life in the alpine meadows will die, covered in snow. In this totality the Van Gujjar thus see the forest—their jungle—as an interdependent system *inclusive* of Van Gujjar and buffaloes.

India's Forest Department, however, has a very different understanding of what a forest is. The British Raj divested forests of existing user rights in Tehri Garhwal, as the area is known, and incorporated them in the imperial Forest Department to be managed for timber production (Chapter 8). The Forest Department trained its officers in scientific forestry to assure a sustainable yield—that is, to regulate timber felling in order not to undermine the reproduction of the stock. The Van Gujjar today require permission from the Forest Department to access their winter pastureland and to lop the trees for fodder.

After India's independence in 1947, a third claim to what a forest is also came to dominate national debate. The commercial plantation model caused rapid biodiversity loss and put the survival of large animals in jeopardy, including the Indian elephant (*Elephas maximus indicus*) and the greater one-horned rhinoceros (*Rhinoceros unicornis*). A vocal Indian environmental movement argued that the forests needed protection from human interference to preserve wildlife. As a result, the Forest Department enclosed large swathes of forest to protect areas in a largely top-down fashion (Shahabuddin 2010). While it is easy to see how the Van Gujjar and the plantation perspectives are at odds with each other, the conservationist discourse has been just as politically trenchant. India's national parks and wildlife sanctuaries often coincide with areas inhabited by forest-dwelling communities who are threatened by displacement (Kabra 2020). Upon returning to their winter pastures in 1992, for example, the Van Gujjar were met by roadblocks preventing them from entering the forest. Their *khol* had been incorporated in the new Rajaji National Park (Gooch 2004).

To think about the power in the Central Himalayas in terms of discourse, we should ask what exactly the forest *is*. Here, three observations stand out. First, the meaning of the "forest" is difficult to establish. Figure 2.3 outlines the contours of three discourses that make claims to the essence of this geographical space. Indeed, the forest only comes into being as a meaningful entity in relation to these larger systems of signs. Second, each discourse derives from a

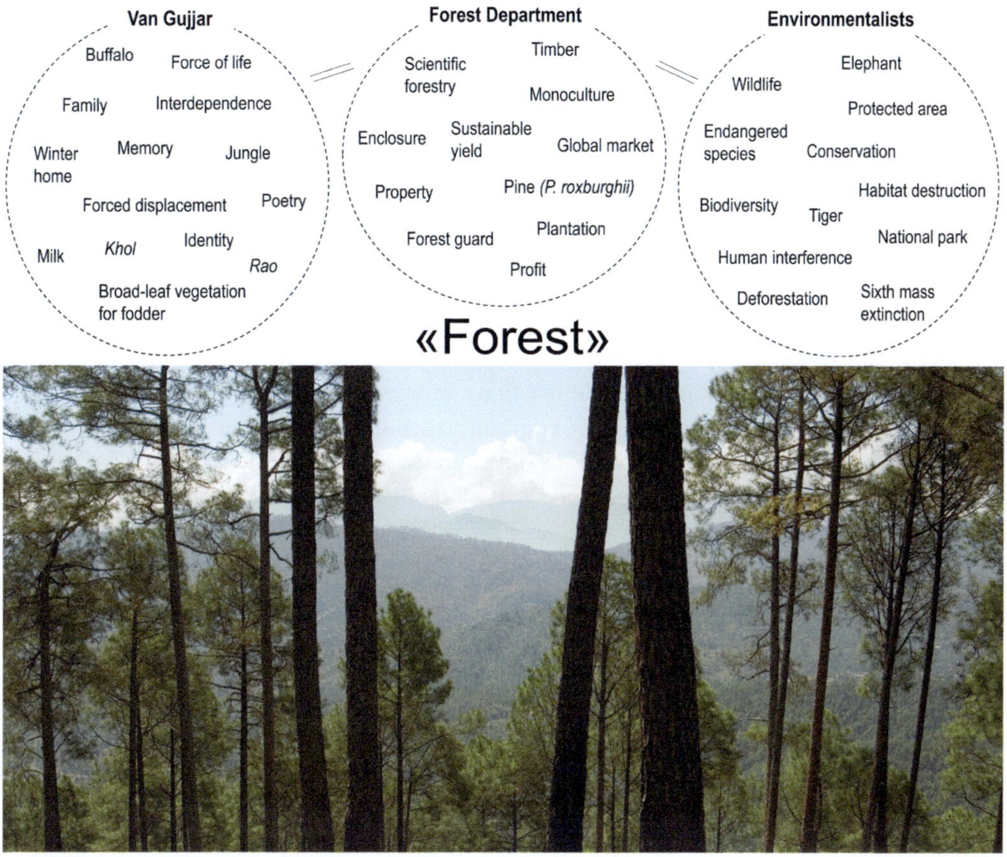

FIGURE 2.3 Three discourses contesting the meaning of the "forest" in the Central Himalayas

It is important to note that this is a very schematic representation. Discourses often draw on each other and are constantly evolving. One human subject can also jump between discourses depending on the context in which they speak. A signifier, such as "forest", that is open to multiple interpretations and whose meaning is contested is referred to as a "floating signifier" in discourse analysis.

more or less institutionalised socio-ecological practice: the Van Gujjar's pastoralism, the Forest Department's timber industry and the environmentalists' conservationist efforts. These practices all seek to transform the forest ecosystem in ways which are cogent and rational within each discourse. As a result, finally, what the forest is and should be is ultimately a contested question that depends on one's situated perspective. Struggle over meaning is deeply political, and power, we might say, operates through discourse.

Working in the early twentieth century, Saussure assumed that there was only one language structure at work in society—stable and unchanging. But an important critique came from the philosopher Michel Foucault, which centred on the relation between discourse and power. In contrast to Saussure, Foucault (2002) studied how a particular discourse becomes dominant at a specific point in time, and how it establishes a form of order that defines what is seen to be normal and true. Truths, Foucault argued, only exist within a discourse as it is impossible to speak from a position outside discourse. A discourse, therefore, exerts power on the mind of a human subject who interprets the world and acts on it based on that worldview. One of Foucault's many contributions was to show how human subjects come into being through discourse, and he examined how individuals recognise themselves as subjects in relation to wider systems of signification. This is an aspect of a Foucauldian perspective on power that we discuss in greater detail in Chapter 8 in terms of "governmentality" and "environmentality" (Agrawal 2005; Li 2007).

In the late 1990s, political ecologists began putting Foucault's ideas to work to study how discourses shape human-environment interactions. However, frequently starting from a more conflictual poststructuralist position than Foucault's, they emphasised how multiple discourses exist at the same time, variously drawing on and contesting each other in a struggle for hegemony of what the world should be (Peet and Watts 1996; Bryant and Bailey 1997; Escobar 1999).

POWER AS KNOWLEDGE

An important feature of Foucault's work, which is not fully captured by the notion of discourse, is the interdependence of power and knowledge. This is a fifth aspect of power that political ecologists engage with. Foucault demonstrated how power presupposes knowledge. For example, it is difficult to envision the modern prison system without criminology, government without national statistics or the colonial timber industry without scientific forestry. In 1735, the Swedish naturalist Carl Linnaeus published *Systema Naturae*, a taxonomic system designed to catalogue all the world's plants, animals and minerals according to one given logic. His was not the first attempt to systematise the natural world, but Linnaeus' approach stood out for its simplicity and elegance. Linnaeus divided all plants into a hierarchy of classes, orders, genera and species based on four visible characteristics: the form, quantity, geometric distribution and size of their reproductive parts. All plants, known and unknown, could be accommodated within the taxonomy, which gave an observer, who saw the world through it, the ability to speak meaningfully about what he or she encountered. In Foucault's (2002: 141, original emphases) words, the classificatory system was "anticipating the possibility of naming; it is the possibility of *seeing* what one will be able to *say*". As such, Linnaeus' taxonomy established a form of order in the environment, but one which was not to be confused with the environment "itself". Rather, it was productive of the environment as a meaningful space.

In *Imperial Eyes*, Mary Louise Pratt (2008: 29) argues that Linnaeus' taxonomy represented "a new form of what one might call planetary consciousness among Europeans". The ambition to observe, categorise and document the natural world in its entirety, based on selected visible traits, reflected a totalising conception of the world. The scientific endeavour coincided with a search for exploitable resources and lands to colonise, and by extension, the need to survey and map these lands in order to govern them (Edney 1997). "One by one", Pratt (2008: 31) writes, "the planet's life forms were to be drawn out of the tangled threads of their life surroundings and rewoven into European-based patterns of global unity and order. The (lettered, male, European) eye that held the system could familiarize ('naturalize') new sites/sights immediately upon contact, by incorporating them into the language of the system".

In the tenth edition of *Systema Naturae*, Linnaeus began classifying the animal kingdom. *Homo sapiens*—humankind—was subdivided into six categories based on physiological and behavioural differences. The European, according to Linnaeus' description, was "Fair, sanguine, brawny; hair yellow, brown, flowing; eyes blue; gentle, acute, inventive. Covered with close vestments. Governed by laws". By contrast, the Asiatic was "Sooty, melancholy, rigid. Hair black; eyes dark; severe, haughty, covetous. Covered with loose garments. Governed by opinions". He singled out Americans, Africans, Wild men and Monsters in even more derogatory terms (Pratt 2008: 32). Linnaean taxonomy, and eighteenth-century Western science more widely, thus produced the modern concept of race, politicising the human body in a hierarchical system with the white European male at its apex. Equipped with this knowledge of the world beyond Europe—the Orient—European travellers were able to understand their place in the world they encountered. Extending Foucault's analysis of power/knowledge, Edward Said (1979: 4–5, original emphases) concludes in his landmark *Orientalism* that "the Orient is not an inert fact of nature. It is not merely *there*, just as the Occident itself is not just *there* either. … [A]s much as the West itself, the Orient is an idea that has a history and a tradition of thought, imagery, and vocabulary that have given it reality and presence in and for the West". Linnaean taxonomy, situated in a broader academic tradition, thus brought a specific form of natural order into being, which legitimised empire and pointed towards particular courses of action.

Political ecologists drawing on Quijano's analysis of modernity/coloniality argue that knowledge is a key area in which the racialised inequalities instituted by European colonialism continue to structure the global economy. Arturo Escobar (2016: 15) shows how there are infinite ways of knowing the world other than a Western understanding; ways of relating to the world "which are far from being exhausted by the Eurocentric experience or reducible to its terms". Yet these other ways of knowing, denoted "epistemologies of the South" (Santos 2014), are repeatedly silenced by Western science, which claims universal applicability. In comparison to Linnaeus' classification of plants, folk taxonomies often group organisms according to habitat, behaviour, morphology, utility or cultural significance. These are no less empirical or rational modes of classification (Malm 2010), but studies in the Linnaean tradition remain widely known as "scientific botany" while non-Western classificatory systems are referred to as "ethnobotany", demoting them to a lesser status as "cultural" curiosities.

By contrast, Linda Tuhiwai Smith (2012) explores research methods with which it is possible to produce academic knowledge *with* rather than *about* otherwise silenced voices. Discussing the relevance of these methods, she concludes that "The globalization of knowledge and Western culture constantly reaffirms the West's view of itself as the centre of legitimate knowledge" (66). Thus, the coloniality of knowledge offers an analytical lens through which we can question whose environmental knowledge gains recognition in development, conservation and research practice. The call to decolonise education and knowledge production is above all a call to understand the world *with* and *from* a subaltern perspective (Daigle and Sundberg 2018; Mignolo and Walsh 2018).

Mining is one contemporary activity which brings the coloniality of knowledge to light, also attesting to the power of discourse. The globalised economy relies on a continuous supply of minerals and energy that is transformed into consumer goods (Chapter 7), and through mining, the environmental load of consumption is geographically displaced from the high street to sites of extraction (Chapter 9)—sites known as the "commodity frontier" or the "resource frontier" (Banoub et al. 2021). Resource frontiers are often embedded in discourses of discovery, wilderness and untapped nature, but even so, they are spaces in which global capital seeks to replace existing systems of knowledge and meaning. On the resource frontier, as Anna Tsing (2003: 5100) argues, landscape elements are wrested "from previous livelihoods and ecologies to turn them into wild resources, available for the industries of the world".

As the polar ice caps melt due to climate change, the Arctic is becoming one of the world's booming resource frontiers (Bennett 2016). In 2014, an Irish-British-Australian business conglomerate sought permission from the Finnish authorities to prospect for kimberlite—a rare mineral known to contain diamonds—in an area called Utsjoki in Finnish or Ohcejohka in Northern Sami. To make its case, the multinational mining company presented a geological map indicating that Utsjoki was "mineral rich" and that a natural "anomaly" existed beneath the mountain Jeagelvárri. Most likely, it was a kimberlite deposit, but only further exploration could tell. Depicting a particular classification of the environment as "mineral rich", the map reinforced a regime of knowledge that rendered the area a site for resource extraction to service the international diamond market.

As Maija Lassila (2018) shows, the proposal sparked local resistance. The majority population in Ohcejohka belong to the indigenous Sami people, and they offered a radically contrasting perspective. Beth Rose Middleton (2015) explains how indigenous cosmologies often emphasise place, rather than time, as a central reference point in the production of knowledge and identity: "place is a storied landscape embodying personal and collective history", she writes (561). The Ohcejohka resistance movement argued that the Jeagelvárri landscape had to be understood holistically in a way such that both humans and nonhumans were part of a wider sociocultural and ancestral network congregating in that place. The landscape was inscribed with memories of reindeer herding and wild-salmon fishing, and knowledge about the environment existed in those continued practices. While the mining company saw the site as a source of profit, the resistance movement saw it as a lived environment, "full of reciprocal, spiritual and identity-making meaning" (Lassila 2018: 6). Thus, the proposed mining enterprise was part of a longer history of colonisation in Sápmi (Chapter 1) in which the region became a resource frontier, violently side-lining traditional Sami livelihoods. The struggle over the meaning of Utsjoki/Ohcejohka exposes a rift between embedded and detached knowledge: it shows how power operates through local and nonlocal discourse, even as the coloniality of knowledge is at work on the resource frontier.

A concept of power as knowledge provides a series of insights to understand human-environment interactions. First, the world needs to be classified and categorised to become meaningful. However, these classificatory systems are not politically innocent seeing that knowledge generates particular forms of order. Second, knowledge as well as discourses are situated in socio-ecological practices and institutions in which they are acted out by human subjects. In doing so, humans shape the world—and themselves—in the image of the classificatory system.

Finally, power is less someone's possession, or a social relation, than a force that operates *through* people in socio-ecological practices. In Foucault's (1980: 98) words, "individuals are the vehicles of power, not its points of application". In this sense, power is not a repressive but a productive phenomenon as the world comes into being through knowledge and discourse. Equipped with this notion of power, the political ecologist's task is not to uncover what the environment "really" looks like but to examine how different social, economic and political interests make claims to the environment. Whose knowledge establishes order and whose discourses are heard? What political, economic and ecological consequences follow? And how are the costs and benefits of those consequences distributed?

CONCLUSION AND LEARNING OUTCOMES

One of the central insights from the science of ecology, which was the starting point of this chapter, is that we cannot fully understand an individual without taking its relations to others into consideration. Human relations are inherently political: human-environment interactions reflect social and material inequalities, uneven patterns of access and conflicting claims to what the environment is. Questions of power must, therefore, be asked in any environmental research. In this chapter, we have identified five different ways in which political ecologists understand power and put it to work as an analytical lens.

Having read this far, we hope that you can account for:

- *Power as a resource*, allowing you to assess how different environmental actors may exert control over a shared environment in contention with each other, for example through coercion, by regulating access to land or through everyday forms of resistance
- *Power as a social relation*, enabling you to analyse how a person's agency and access to resources are influenced by multiple intersecting social constructs, such as class, gender and race
- *Power as ideology*, allowing you to examine how material inequalities are normalised through dominant cultural constructs that make them appear as self-evident and natural
- *Power as discourse*, enabling you to analyse how the environment attains meaning through discourses rooted in socio-ecological practices, and how claims to what the environment is and should be ultimately are contested
- *Power as knowledge*, allowing you to investigate how power and knowledge presuppose each other and how European knowledge systems systematically marginalise non-European epistemologies at the resource frontier

FURTHER READING

Quijano, Aníbal. 2000. "Coloniality of Power, Eurocentrism, and Latin America." *Nepantla: Views from South* 1(3): 533–580.

Quijano's work is foundational to discussions on modernity/coloniality, and this text is essential reading if you want to gain a more in-depth understanding of how power can be seen to operate on a global scale, in a "world system" characterised by unequal class and race relations.

Svarstad, Hanne, Tor A. Benjaminsen, and Ragnhild Overå. 2018. "Power Theories in Political Ecology." *Journal of Political Ecology* 25: 350–363.

Svarstad et al.'s paper is an excellent text to read after this chapter as it adds further layers to how the concept of power has been theorised by political ecologists. Discussing "actor-oriented" and "neo-Marxist" approaches, it also introduces a collection of papers that each engage with Foucauldian power theories.

BIBLIOGRAPHY

Agrawal, Arun. 2005. *Environmentality: Technologies of Government and the Making of Subjects*. Durham, NC: Duke University Press.

Appelbaum, Nancy P., Anne S. Macpherson, and Karin Alejandra Rosemblatt. 2003. "Racial Nations." In *Race and Nation in Modern Latin America*, edited by Nancy P. Appelbaum, Anne S. Macpherson, and Karin Alejandra Rosemblatt, 1–31. Chapel Hill, NC: University of North Carolina Press.

Asiyanbi, Adeniyi P. 2016. "A Political Ecology of REDD+: Property Rights, Militarised Protectionism, and Carbonised Exclusion in Cross River." *Geoforum* 77: 146–156.

Aston, T. H., and C. H. E. Philpin, eds. 1985. *The Brenner Debate: Agrarian Class Structure and Economic Development in Pre-Industrial Europe*. Cambridge, UK: Cambridge University Press.

Banoub, Daniel, Gavin Bridge, Beatriz Bustos, Irmak Ertör, Marien González-Hidalgo, and Julie Ann de los Reyes. 2021. "Industrial Dynamics on the Commodity Frontier: Managing Time, Space and Form in Mining, Tree Plantations and Intensive Aquaculture." *Environment and Planning E: Nature and Space* 4(4): 1533–1559.

Bennett, Mia M. 2016. "Discursive, Material, Vertical, and Extensive Dimensions of Post-Cold War Arctic Resource Extraction." *Polar Geography* 39(4): 258–273.

Blaikie, Piers. 1985. *The Political Economy of Soil Erosion in Developing Countries*. London, UK: Routledge.

Bryant, Raymond L., and Sinéad Bailey. 1997. *Third World Political Ecology*. London, UK: Routledge.

Butler, Judith. 1990. *Gender Trouble: Feminism and the Subversion of Identity*. London, UK: Routledge.

Crenshaw, Kimberle. 1991. "Mapping the Margins: Intersectionality, Identity Politics, and Violence against Women of Color." *Stanford Law Review* 43(6): 1241–1299.

Dahl, Robert A. 1957. "The Concept of Power." *Behavioral Science* 2(3): 201–215.

Daigle, Michelle, and Juanita Sundberg. 2018. "From Where We Stand: Unsettling Geographical Knowledges in the Classroom." *Transactions of the Institute of British Geographers* 42(3): 338–341.

Dove, Michael R., and Daniel M. Kammen. 2015. *Science, Society and the Environment: Applying Anthropology and Physics to Sustainability*. London, UK: Routledge.

Duffy, Rosaleen, Freya A. V. St John, Bram Büscher, and Dan Brockington. 2016. "Toward a New Understanding of the Links between Poverty and Illegal Wildlife Hunting." *Conservation Biology* 30(1): 14–22.

Duncan, James S. 1989. "The Power of Place in Kandy, Sri Lanka: 1780–1980." In *The Power of Place: Bringing Together Geographical and Sociological Imaginations*, edited by John A. Agnew and James S. Duncan, 185–201. London, UK: Routledge.

Edney, Matthew H. 1997. *Mapping an Empire: The Geographical Construction of British India, 1765–1843*. Chicago, IL: University of Chicago Press.

Ekers, Michael, Gillian Hart, Stefan Kipfer, and Alex Loftus, eds. 2013. *Gramsci: Space, Nature, Politics*. Oxford, UK: Wiley-Blackwell.

Escobar, Arturo. 1995. *Encountering Development: The Making and Unmaking of the Third World*. Princeton, NJ: Princeton University Press.

Escobar, Arturo. 1999. "After Nature: Steps to an Antiessentialist Political Ecology." *Current Anthropology* 40(1): 1–30.

Escobar, Arturo. 2016. "Thinking-Feeling with the Earth: Territorial Struggles and the Ontological Dimension of the Epistemologies of the South." *AIBR. Revista de Antropología Iberoamericana* 11(1): 11–32.

Estes, Nick. 2019. *Our History Is the Future: Standing Rock Versus the Dakota Access Pipeline, and the Long Tradition of Indigenous Resistance*. London, UK: Verso.

Fletcher, Robert, Wolfram Dressler, and Bram Büscher. 2015. "Nature™ Inc.: Nature as Neoliberal Capitalist Imaginary." In *International Handbook of Political Ecology*, edited by Raymond L. Bryant, 359–372. Cheltenham, UK: Edward Elgar.

Foucault, Michel. 1980. "Two Lectures." In *Power/Knowledge: Selected Interviews and Other Writings, 1972–1977*, edited by Colin Gordon, 78–108. New York, NY: Pantheon.

Foucault, Michel. 2002 (1966). *The Order of Things: An Archaeology of the Human Sciences*. London, UK: Routledge.

Galeano, Eduardo, 1997 (1973). *Open Veins of Latin America*. New York: Monthly Review Press.

Gilio-Whitaker, Dina. 2019. *As Long as Grass Grows: The Indigenous Fight for Environmental Justice from Colonization to Standing Rock*. Boston, MA: Beacon Press.

Gooch, Pernille. 2004. "Van Gujjar: The Persistent Forest Pastoralists." *Nomadic Peoples (N.S.)* 8(2): 125–135.

Gramsci, Antonio. 1971. *Selections from the Prison Notebooks*. London, UK: Lawrence and Wishart.

Grosfoguel, Ramón. 2007. "The Epistemic Decolonial Turn." *Cultural Studies* 21(2–3): 211–223.

Harcourt, Wendy, and Ingrid L. Nelson, eds. 2015. *Practising Feminist Political Ecologies: Moving beyond the "Green Economy*. London, UK: Zed.

Hart, Gillian. 2013. "Gramsci, Geography, and the Languages of Populism"." In *Gramsci: Space, Nature, Politics*, edited by Michael Ekers, Gillian Hart, Stefan Kipfer, and Alex Loftus, 301–320. Oxford, UK: Wiley-Blackwell.

Harvey, David. 1974. "Population, Resources, and the Ideology of Science." *Economic Geography* 50(3): 256–277.

Isaac, Jeffrey C. 1987. "Beyond the Three Faces of Power: A Realist Critique." *Polity* 20(1): 4–31.

Ives, Peter. 2004. *Language and Hegemony in Gramsci*. London, UK: Pluto Press.

Kabra, Asmita. 2020. "Caste in Stone? Exploring Caste and Class Dimensions of Conservation Displacement in Central India." *Journal of Contemporary Asia* 50(5): 785–805.

Karriem, Abdulrazak. 2013. "Space, Ecology and Politics in the Praxis of the Brazilian Landless Movement." In *Gramsci: Space, Nature, Politics*, edited by Michael Ekers, Gillian Hart, Stefan Kipfer, and Alex Loftus, 142–160. Oxford, UK: Wiley-Blackwell.

Lassila, Maija M. 2018. "Mapping Mineral Resources in a Living Land: Sami Mining Resistance in Ohcejohka, Northern Finland." *Geoforum* 96: 1–9.

Lefebvre, Henri. 1982 (1968). *The Sociology of Marx*. Translated by Norbert Guterman. New York, NY: Columbia University Press.

Leff, Enrique. 2015. "Encountering Political Ecology: Epistemology and Emancipation." In *International Handbook of Political Ecology*, edited by Raymond L. Bryant, 44–56. Cheltenham, UK: Edward Elgar.

Li, Tania Murray. 2007. *The Will to Improve: Governmentality, Development, and the Practice of Politics*. Durham, NC: Duke University Press.

Loftus, Alex. 2013. "Gramsci, Nature, and the Philosophy of Praxis." In *Gramsci: Space, Nature, Politics*, edited by Michael Ekers, Gillian Hart, Stefan Kipfer, and Alex Loftus, 178–196. Oxford, UK: Wiley-Blackwell.

Loomba, Ania. 2015. *Colonialism/Postcolonialism*, 3rd ed. London, UK: Routledge.

Lukes, Steve. 2005 (1974). *Power: A Radical View*. New York, NY: Palgrave Macmillan.

Malm, Thomas. 2010. "Why Is the Shark Not an Animal? On the Division of Life-Form Categories in Oceania." *SPC Traditional Marine Resource Management and Knowledge Information Bulletin* 27: 17–22.

Marx, Karl. 1976 (1867). *Capital: A Critique of Political Economy*, vol. 1. London, UK: Penguin.

Middleton, Beth Rose. 2015. "*Jahát Jatítotòdom*: Toward an Indigenous Political Ecology." In *International Handbook of Political Ecology*, edited by Raymond L. Bryant, 561–576. Cheltenham, UK: Edward Elgar.

Mignolo, Walter D., and Catherine E. Walsh. 2018. *On Decoloniality: Concepts, Analytics, Praxis*. Durham, NC: Duke University Press.

Mollett, Sharlene, and Caroline Faria. 2013. "Messing with Gender in Feminist Political Ecology." *Geoforum* 45: 116–125.

Moore, Donald. 1996. "Marxism, Culture and Political Ecology: Environmental Struggles in Zimbabwe's Eastern Highlands." In *Liberation Ecologies: Environment, Development, Social Movements*, edited by Richard Peet and Michael Watts, 125–147. London, UK: Routledge.

Nietschmann, Bernard. 1973. *Between Land and Water: The Subsistence Ecology of the Miskito Indians, Eastern Nicaragua*. New York, NY: Seminar Press.

Nightingale, Andrea J. 2011. "Bounding Difference: Intersectionality and the Material Production of Gender, Caste, Class and Environment in Nepal." *Geoforum* 42: 153–162.

Paulson, Susan, Lisa L. Gezon, and Michael Watts. 2005. "Politics, Ecologies, Genealogies." In *Political Ecology across Spaces, Scales, and Social Groups*, edited by Susan Paulson and Lisa L. Gezon, 17–37. New Brunswick, NJ: Rutgers University Press.

Peet, Richard, and Michael Watts. 1996. "Liberation Ecology: Development, Sustainability, and Environment in an Age of Market Triumphalism." In *Liberation Ecologies: Environment, Development, Social Movements*, edited by Richard Peet and Michael Watts, 1–45. London, UK: Routledge.

Peluso, Nancy Lee. 2012. "What's Nature Got to Do with It? A Situated Historical Perspective on Socio-Natural Commodities." *Development & Change* 43(1): 79–104.

Pratt, Mary Louise. 2008. *Imperial Eyes: Travel Writing and Transculturation*, 2nd ed. London, UK: Routledge.

Quijano, Aníbal. 2000. "Coloniality of Power, Eurocentrism, and Latin America." *Nepantla: Views from South* 1(3): 533–580.

Rangarajan, Mahesh, and K. Sivaramakrishnan, eds. 2012. *India's Environmental History: Colonialism, Modernity, and the Nation*, vol 2. Ranikhet, India: Permanent Black.

Rappaport, Roy. 1968. *Pigs for the Ancestors: Ritual and Ecology of a New Guinea People*. New Haven, CT: Yale University Press.

Ribot, Jesse C., and Nancy Lee Peluso. 2003. "A Theory of Access." *Rural Sociology* 68(2): 153–181.

Robbins, Paul. 2000. "The Rotten Institution: Corruption in Natural Resource Management." *Political Geography* 19(4): 423–443.

Robbins, Paul. 2020. *Political Ecology: A Critical Introduction*, 3rd ed. Oxford, UK: Wiley-Blackwell.

Rocheleau, Dianne, Barbara Thomas-Slayter, and Esther Wangari, eds. 1996. *Feminist Political Ecology: Global Issues and Local Experiences*. London, UK: Routledge.

Said, Edward W. 1979. *Orientalism: Western Conceptions of the Orient*. New York, NY: Vintage Books.

Santos, Boaventura de Sousa. 2014. *Epistemologies of the South: Justice against Epistemicide*. London, UK: Routledge.

Schmink, Marianne, and Charles H. Wood. 1992. *Contested Frontiers in Amazonia*. New York, NY: Columbia University Press.

Scott, James C. 1985. *Weapons of the Weak: Everyday Forms of Peasant Resistance*. New Haven, CT: Yale University Press.

Shahabuddin, Ghazala. 2010. *Conservation at the Crossroads: Science, Society, and the Future of India's Wildlife*. New Delhi, India: Orient Blackswan.

Sharma, Mukul. 2018. *Caste and Nature: Dalits and Indian Environmental Politics*. New Delhi, India: Oxford University Press.

Shiva, Vandana. 2014. *The Vandana Shiva Reader*. Lexington, KY: University Press of Kentucky.

Smith, Linda Tuhiwai. 2012. *Decolonizing Methodologies: Research and Indigenous Peoples*, 2nd ed. London, UK: Zed.

Sultana, Farhana. 2021. "Political Ecology 1: From Margins to Center." *Progress in Human Geography* 45(1): 156–165.

Sundberg, Juanita. 2008. "Placing Race in Environmental Justice Research in Latin America." *Society & Natural Resources* 21(7): 569–582.

Sundberg, Juanita. 2017. "Feminist Political Ecology." In *International Encyclopedia of Geography: People, the Earth, Environment and Technology*, edited by Douglas Richardson, n.p. New York, NY: Wiley-Blackwell.

Svarstad, Hanne, Tor A. Benjaminsen, and Ragnhild Overå. 2018. "Power Theories in Political Ecology." *Journal of Political Ecology* 25: 350–363.

Thompson, Edward P. 1980 (1963). *The Making of the English Working Class*. London, UK: Penguin.

Truelove, Yaffa. 2019. "Gray Zones: The Everyday Practices and Governance of Water beyond the Network." *Annals of the American Association of Geographers* 109(6): 1758–1774.

Tsing, Anna L. 2003. "Natural Resources and Capitalist Frontiers." *Economic & Political Weekly* 38(48): 5100–5106.

VeneKlasen, Lisa, and Valerie Miller. 2002. *A New Weave of Power, People & Politics: The Action Guide for Advocacy and Citizen Participation*. Oklahoma City, OK: World Neighbors.

von Schnitzler, Antina. 2013. "Traveling Technologies: Infrastructure, Ethical Regimes, and the Materiality of Politics in South Africa." *Cultural Anthropology* 28(4): 670–693.

Watts, Michael J. 2013 (1983). *Silent Violence: Food, Famine, and Peasantry in Northern Nigeria*. Athens, GA: University of Georgia Press.

Williams, Raymond. 1983. *Keywords: A Vocabulary of Culture and Society*. Oxford, UK: Oxford University Press.

Wisner, Ben, Piers Blaikie, Terry Cannon, and Ian Davis. 2004. *At Risk: Natural Hazards, People's Vulnerability and Disasters*, 2nd ed. London, UK: Routledge.

Wolf, Eric R. 1982. *Europe and the People without History*. Berkeley, CA: University of California Press.

Yenneti, Komali, Rosie Day, and Oleg Golubchikov. 2016. "Spatial Justice and the Land Politics of Renewables: Dispossessing Vulnerable Communities through Solar Energy Mega-Projects." *Geoforum* 76: 90–99.

CHAPTER 3

Scale

In its different traditions, political ecology has taken questions of scale fundamentally seriously. Within its Anglo-American guise, as we discussed in Chapter 1, the field is seen to have emanated from a discussion concerning the scale at which to explain environmental change. *Global* environmental change has subsequently generated a whole new set of concerns, and many political ecologists remain deeply concerned with scalar questions: how does the local articulate global processes and how does the global find expression in lived local realities? In this chapter, we introduce scale as a key concept. Like power, it is a term with contested meaning, and while it is central to any study in ecology (Sayre 2015), long-standing debates around scale have unfolded involving political ecologists and human geographers in particular. In the early 2000s, some even called for the concept's abandonment. In the following, we unpack the notion of scale to highlight the utility of this concept for political ecology, and in doing so, we pay extra attention to the crucial contributions of political ecologists to these debates.

A CALIFORNIAN CRISIS?

Manufacturing and a large defence industry, but also agriculture, have long dominated the State of California. Dependent on well-drained soils and a warm climate, cotton grows particularly well in the Californian valleys. During the mid-twentieth century, an economic paradigm known as Keynesianism, holding that government intervention could help to stave off economic crises through targeted fiscal policy, had taken a distinct form in the state. Federal spending was targeted on defence to generate employment on the assembly lines producing armaments. However, as elsewhere in the world economy, California entered a period of "stagflation" in the 1970s, thereby generating a serious crisis for a growth economy. With this economic downturn, the Keynesian consensus was itself thrown into crisis. The decline in the Californian model of "military Keynesianism" coincided with a broader process of de-industrialisation, and while unemployment soared in manufacturing, a drought crippled parts of the state's agricultural sector. Cotton production was particularly affected.

The division of labour—whether in agriculture, industry or the service economy—has historically been shaped by deeply racialised differences in California. The relative force of de-industrialisation, along with suppressed agricultural employment and wages, was

DOI: 10.4324/9781003095521-3

therefore felt differently among social groups. In *Golden Gulag*, Ruth Wilson Gilmore (2007) analyses the massive growth in California's prison population immediately after the crisis of the 1970s. In doing so, she presents a detailed account of how political economy, environmental processes and changing social identities converge in particular places at particular times. Although often implicitly, Gilmore's analysis tells us much about the ways in which scale operates. At the global, federal, state, local and bodily scales, different economic and environmental processes were patterning a distinct political ecology within and across the region. For example, Gilmore shows how large surpluses of finance capital, labour, land and state capacity were unable to find markets for investment, regionally and globally. Investing in prisons became a solution to this political ecological crisis. The "prison fix", as Gilmore calls it, was a distinctly Californian response to some of the contradictory processes unfolding within the state.

Simultaneously, the response from activists mobilising to frustrate the prison fix was also a scalar one. As Wendy Cheng and Rashad Shabazz (2015: 5) argue:

> anti-prison activists started asking questions from a "local" level but came to expand their geographic scale of analysis statewide, in order to understand the linkages between the inner-city areas from which most prisoners came and the rural towns in which prisons were sited. Being attentive to geographic scale as well as to various intellectual scales of analysis is essential to understanding social [and environmental] problems, since it helps us understand how the specificities of local conditions are rooted in more expansive circuits of power that operate across scale.

There are three important ways in which scale operates in these analyses:

1. Scale provides a conceptual framework allowing us to make sense of the patterning of social, ecological and economic processes. In other words, it is an analytical tool.
2. Scale refers to the patterns themselves and the ways in which they converge at distinct levels. It is something that exists "out there".
3. Scale is a site of struggle that opens up the possibility for forging fundamentally new human-environment relations on the grounds of the old.

We can look more closely at each meaning of scale through the Californian case. First, in analysing the Californian landscape, it is useful to *think* in terms of different scales. Gilmore abstracts from the complex realities of California to focus on processes operating in the global economy down to the lives of individual, racialised people and their "marked" or "unmarked" bodies. In this instance, scale is an epistemological concept, offering an analytical toolkit for making sense of political ecology.

Second, key social, economic and environmental processes, such as the turn away from military Keynesianism and the emergence of drought conditions, converged in California to produce distinct patterns in the landscape. Surpluses of land, labour and capital emerged that could only be resolved through "the prison fix". This fix was itself a distinctive scalar response to broader changes in the world economy. In this case, scale is an actually existing ontological phenomenon.

Third, having observed scalar patterns in the landscape and produced an analysis that seeks to interpret them, attention to scale also enabled activists to engage in concrete struggles that could bring about social and environmental change. In Gilmore's work, the social movements calling for prison abolition sought change beyond (while of course including) freedoms for individual people, and as a movement, prison abolition worked across a range of different scales. In Gilmore's words, "Abolition is deliberately everything-ist; it's about the entirety of human-environmental relations" (quoted in Kushner 2021). Just as "political ecology" can be understood in terms of epistemology, ontology and praxis (Chapter 1), then, so can the key concept of scale (Table 3.1).

TABLE 3.1 Multiple meanings and uses of scale

Scale as...	Focus	Aspect	Consists of	Example
Analytical concept	Epistemology	Size	Temporal scale	Is a geological age, a season or a day the relevant temporal unit of observation?
			Observational scale	Is a home, a city or an integrated urban-rural system the relevant spatial unit of observation?
			Cartographic scale	Should the world be represented covering a large extent, resulting in lesser detail, or a small extent, resulting greater detail?
		Level	Classificatory or institutional hierarchy	Ecosystem Community Population Species Organism International National Regional Municipal Local
			A chain of explanation	Explaining local experiences by contextualising them via their relations to larger scalar levels
Converging patterns	Ontology	Relation	Material processes interacting across and between scales	In California: stagflation, drought and a racialised division of labour coming together with effects in everyday life
				In Spain: the physical, institutional and symbolic interconnection of several river basins
Struggle and process	Praxis	Scale jumping	Local protests or actions joining together so that a change in quantity results in a change in quality	Greta Thunberg's one-person protest becoming a global movement for climate justice

FROM GEOGRAPHICAL TO TEMPORAL SCALE

While *Golden Gulag* suggests different ways of understanding geographical scale, political ecologists also regularly confront temporal scales. Temporal scales are perhaps more widely acknowledged than spatial scales. Indeed, we are often trained to think in terms of "pre-historic times", "the Middle Ages" and "the colonial era"—terms that tell us as much about who is writing history as they do about history itself. As Gabrielle Hecht (2018: 114) argues, scale is "about categories: what they reveal or hide, the ways in which they do (or do not) nest". Ecological processes can also be seen to exhibit distinct temporal patterns. We might, for example, think about those weather patterns or climatic abstractions that we give names to—Summer, El Niño Southern Oscillation and global warming. Epistemologically, geologists and earth system scientists will periodise time to try to better understand these patterns. The Holocene, Pleistocene and Pliocene are just three of the more recent periods identified in the geological calendar. For political ecologists, the temporal scales of socio-ecological change become even more interesting when we identify the processes and relationships that give rise to them. For example, a critical study of climate change necessarily involves identifying spatial and temporal shifts, naming the processes that are responsible for them and seeking to ally with struggles that might produce better socio-ecological outcomes.

We saw one of the most obvious examples of such work in Chapter 1 when we introduced "the Anthropocene". This concept frames our current epoch as the "age of the human", which puts emphasis on social and biophysical processes that work on a very long geological temporal scale. Many political ecologists remain rightly agnostic about the Anthropocene concept for the way it identifies an abstract "human" to be responsible for our current condition when so evidently it is the world's most affluent citizens who contribute the most to carbon emissions and environmental change (Malm and Hornborg 2014; Davis et al. 2019; Wiedmann et al. 2020). However, the Anthropocene concept demonstrates how different ways of thinking about temporal and geographical scale coalesce. Indeed, the Anthropocene provides something of a challenge to the scales used by many geologists, earth system scientists, historians and critical theorists in their research. For Hecht (2018: 113), the Anthropocene "asserts scalar enormity".

Paul Crutzen (2002), who coined the term, initially traced the beginning of the Anthropocene to the latter part of the 1700s, a period in which increased levels of carbon dioxide (CO_2) and methane in the atmosphere began to leave a detectable trace within polar ice. As Crutzen went on to note, the date corresponded "roughly" with James Watt's invention of the steam engine in 1784 (Chapter 7). Nevertheless, subsequent scholars have questioned Crutzen's 1784 date. In one of the more interesting studies, Simon Lewis and Mark Maslin (2015) argue that the start of the epoch should be dated to either 1610 or 1964. The latter date refers to the so-called "bomb spike" in the carbon-dating record, which resulted from the planetary dispersal of radioactive ash from nuclear testing. The former date, the so-called "Orbis spike", denotes the massive loss of life caused by European colonisers in the Americas. The death of around 100 million people, through murder, disease and the destruction of livelihoods, resulted in a dramatic decline in CO_2 levels around the world. When vast numbers of native farmers died, forests grew back and the CO_2 levels declined. The seventeenth century, then, witnessed a massive expansion in the scale at which human-environment relations operated through imperial violence. Arguably, this shift in the scale of human influence on the biosphere also necessitates a conceptual shift in the spatial and temporal scales at which environmental questions are to be analysed. (For a good summary of the estimates of population decline and the reasons for it, see Newson 1993, and for an accessible environmental history covering some of these themes, see Marks 2019.)

If we combine the discussion of Gilmore's work on the prison fix with that of the Anthropocene, we can identify at least four different themes that political ecologists need to consider when discussing scale. First, scale is not something that is permanently fixed. While political ecologists might focus their studies on a single village, a city, a country or a land mass, and while there may be good analytical reasons for focusing on that "site", we should always remain attentive to processes that exceed the scale at which we are working and that produce and reproduce that scale. In the case of Lewis and Maslin's (2015) analysis, the processes of colonisation and neo-colonisation are just two examples of processes of rescaling that have had dramatic implications for human and nonhuman life, as well as for the environments of which they are a part. The decline of Keynesianism in the case of *Golden Gulag* is another case of multiple processes exceeding the local site.

Second, and tempering this earlier point slightly, environmental processes really do cohere at certain moments, in particular places, and around certain temporal and spatial scales. Drought in California, for example, made cotton production in the state prohibitively expensive, which led to a surplus of land and labour. Furthermore, calls for governments to acknowledge the climate emergency on international, national and city levels have been bolstered by evidence of the crossing of key planetary boundaries (Rockström et al. 2009; Steffen et al. 2015). In seeking to appeal to policymakers, Crutzen and others show that being aware of and analysing the breaching of temporal and spatial scales is crucially important for effective environmental governance.

Third, to utilise scale as an analytical tool is not simply about recognising patterns that already exist in the world. Beyond recognition, the lenses political ecologists use—and the granularity of those lenses—make a difference to what researchers can and cannot see. In other words, there is a politics associated with knowledge production linked to questions of scale. To use a very crude example, an analysis of austerity in the United Kingdom that focuses on a uniformly rich neighbourhood, such as a gated community in North London, is unlikely to be able to identify many of the key effects of what has been a national set of policies. Scale is, therefore, also about an epistemological choice—a thought process of deciding where to focus one's study and how to capture the relationships that need to be researched (Cederlöf and Hornborg 2021).

Finally, the struggle to define scale is a political one that opens the possibility for new forms of solidarity and new forms of socio-ecological organisation to develop. The ability to reconfigure scale is brilliantly exemplified in Gilmore's analysis of the movement Mothers Reclaiming Our Children (Mothers ROC), one of several activist groups covered within her work. Through the principles "Educate, Organize, Empower", Mothers ROC emerged in response to the prison crisis in California. The group was able to "jump scale", as Neil Smith (2007) might refer to it, identifying with a much broader set of struggles and becoming the kind of "everything-ist" movement that Gilmore associates with abolitionism. Indeed, Gilmore builds on the work of Antonio Gramsci (Chapter 2) to interpret Mothers ROC's ability to move from the geographically situated activity of fighting California's prison fix to also embrace broader political ambitions—a movement of both theory and practice.

TRACING THE SCALE DEBATES IN POLITICAL ECOLOGY: SCALE AS SIZE? SCALE AS LEVEL? SCALE AS RELATION?

In his review of the debates over scale in political ecology, Nathan Sayre (2015) summarises the main ways in which scale tends to be discussed. One definition of scale, he argues, is *scale as size*. For example, we can talk about the scale on a ruler, in which case "scale" is a measure. Scale as size or measure is also how the concept traditionally is understood by cartographers. The Ordnance Survey maps that cover the United Kingdom are generally available in a 1:50,000 scale or a 1:25,000 scale, with 1 unit on the map measuring 50,000 times or 25,000 times bigger in the world that it seeks to chart. When Britain set out to map its empire in the eighteenth and nineteenth centuries, and thereby to construct "the empire" as an object of knowledge (Chapter 2), the question of the scale at which to map the imperial landscape was also subject to heated discussion (Edney 1997). Such uses of scale as size should be distinguished from those that are concerned with the scale at which observations take place. In this case, scale is about the resolution, granularity, or extent of a study. For example, we might speak about a large-scale study of biodiversity loss across the United Kingdom, but to represent this study area visually would require a smaller scale map than if we were to focus on a city alone.

As opposed to size, *scale as level* refers to the classificatory processes through which researchers might group certain processes. Sayre notes that within the science of ecology, scale is often referred to at the level of the "organism" or the "community". Many studies in political ecology, in comparison, work with an understanding of scale in terms of nested institutional levels, such as the local, municipal, regional, national, and international levels of governance. As we return to below, the notion of scale as level was particularly important for the early development of political ecology in the Anglo-American tradition when scholars applied a "chain of explanation" to environmental issues.

Most interesting for Sayre, however, is the use of *scale as relation*, for "only as a relation", he argues, "does scale acquire the power ascribed to it" (2009: 96). To understand scale as relation, it is helpful to consider Erik Swyngedouw's reinterpretation of globalisation as "glocalisation". The 1990s was a decade when debates around globalisation reached a zenith. These debates very often pivoted around whether the scale of the nation state was fundamentally endangered or, paradoxically, whether it was being strengthened by processes of globalisation. Applying a dialectical framework, Swyngedouw (2004) argued that what was actually taking place was a reconfiguration of the role of the nation state *in relation to* other scales—both global and local. The "global" and the "local" were not replacing the "national"; instead, the relations between these scales—as well as people's abilities to act on them—were changing. To isolate one scale outside of these reconfigurations was to produce a one-sided and flawed interpretation, failing to capture the complexity of rescaling as a process. Responding to the question of whether the world was becoming more globalised or less, Swyngedouw argued that it was becoming both.

SCALE AS A CHAIN OF EXPLANATION

Such debates over the patterning of socio-ecological processes and the appropriate scale at which to analyse them are intimately bound up in the development of political ecology. Indeed, as we noted earlier, one of the field's most widely accepted origin stories revolves around a debate over the most appropriate scale for research. According to this origin story, studies of cultural ecology had become too rooted in local descriptions to interpret how global processes were transforming its sites of study. Bernard Nietschmann's (1973) work on the Miskito of Nicaragua is often viewed as part of a paradigm shift from cultural to political ecology within the Anglo-American tradition. Following the lead of other cultural ecologists, Nietschmann at first sought to understand how human cultural practices and forms of social organisation maintained the stability of ecosystems. In such analyses, the boundary of a social system tended to map neatly on to the boundary of an ecosystem (e.g. Rappaport 1968; Barth 2008). However, as Nietschmann quickly recognised, by the late 1960s and early 1970s, the subsistence ecology of the Miskito peoples was being transformed. Crucially, the sea turtles they fished for had become exchangeable commodities. Nietschmann noted that the developing market economy did not seem to serve as an adaptive mechanism keeping an ecosystem in equilibrium; instead, it was transforming the lives and ecologies of the Miskito coast in a new direction (Figure 3.1).

While cultural ecologists had focused their studies firmly on one site, such as the Miskito coast, Nietschmann's study pointed to the limits of an approach bounded at a local scale. Through the process of commodification, the practices of the Miskito were being integrated in wider spheres of exchange, acting at larger scales. As we suggested in Chapter 1, subsequent works by Anglo-American scholars like Nancy Lee Peluso (1992), Piers Blaikie (1985), Susanna Hecht (1985) and Michael Watts (2013) went on to challenge the foundations of cultural ecology. For this emerging group of political ecologists, the bounded systems studied in cultural ecology, in which societies mapped neatly on to ecosystems, were in the process of being transformed by a developing and deepening capitalist mode of production. In the twentieth century, this mode of production was clearly operating on a global scale. Thus, cultural ecology's focus on "the site" had become too autarkic and inward-looking; too constrained by the models and frameworks it worked with to be able to research something so transformative as globalised capitalism (Neumann 2005; Dove 2006).

With the site apparently no longer viable as a bounded unit of study, political ecologists were forced to think differently about the scales at which their research should be focused. Piers Blaikie and Harold Brookfield's collection *Land Degradation and Society* (1987), therefore, proposed a "chain of explanation" that might form the basis for future studies. This chain of explanation was based around a nested hierarchy of scales (thereby coinciding with Sayre's 2015 definition of scale as "level") in which the researcher should start out in a local context and then move from the person to the household, the village, the region, the state and finally the world (Chapter 8). Blaikie and Brookfield's approach makes certain intuitive sense: the changes

FIGURE 3.1 Green sea turtles (*Chelonia mydas*) off the coast of Nicaragua

The International Union for the Conservation of Nature (IUCN) today lists the green sea turtle as an endangered species. A cultural ecological explanation of their decimating numbers would have focused on human practices in a circumscribed study area, but Nietschmann found that the fishing practices of the Miskito had to be understood in relation to the globalising market economy of which they were becoming a part.

Photo by iStock.com/andyKRAKOVSKI

witnessed by individuals on the Miskito coast were influenced by village-level decisions and national policies in Nicaragua, which themselves responded to broader geopolitical shifts, not least the rapid development of a capitalist economy. Hence, the chain of explanation provided a basis from which political ecologists could think about patterns of environmental change, as well as how best to research the underlying causal processes shaping those patterns.

For an incipient subdiscipline, the chain of explanation provided political ecology with a framework that could be productively applied in subsequent analyses. Nevertheless, as many have later pointed out, there were risks associated with such a framework if it was applied too rigidly. First, if scale is interpreted as a fixed category of nested levels, it is easy to overlook processes that cut across the different levels as well as how the relations between scales—and thus the abilities of people to act on them—are reconfigured in times of environmental and political economic change (Swyngedouw 2004). Second, the scalar levels defined by the framework may begin to dictate the type of analysis being conducted. In this case, the framework defines the processes that are considered important, opening the research up to the kind of criticism advanced by Andrew Vayda and Bradley Walters (1999) against political ecology (Chapter 1). This reversal of roles, in which a fixed analytical framework takes precedence over a more open-ended empiricism, is also what has been described as a spatial (or indeed scalar) fetishism.

SPATIAL FETISHISM: METHODOLOGICAL NATIONALISM AND CITYISM

The critique of "spatial fetishism" is inspired by a famous section in the first chapter of Karl Marx's *Capital*. Marx (1976) writes about how commodities tend to exert a power over the people who made them. A table can be transformed into a commodity which can then be traded for other commodities. Through such a process, Marx states, the table, while still an object created by humans, appears to have a life of its own:

> the table continues to be wood, an ordinary sensuous thing. But as soon as it emerges as a commodity, it changes into a thing which transcends sensuousness. It not only stands with its feet on the ground, but in relation to all other commodities, it stands on its head, and evolves out of its wooden brain grotesque ideas far more wonderful than if it were to begin dancing of its own free will.
>
> (163–164)

Just as a fetish—to outsiders, at least—is a thing imbued with mystical powers by those who constructed it, so a commodity can exert mystical powers over the worker who produced it (see Chapter 9 for a fuller discussion of commodity fetishism). Marx takes these powers seriously: the commodity fetish is more like a mirage (an objective phenomenon) than an illusion (a subjective phenomenon), but more importantly, he takes a genuine interest in the various social relations that enable this inversion of people and things.

The same kind of approach has been adopted by those who criticise the fetishisation of certain spatial scales. There is nothing natural, fixed, innate or given about a particular scale. Nevertheless, as with Marx's table, there are times when these scales can begin to exert a peculiar power of their own over the humans that created them. For example, the nation state has had a relatively brief history as a means of organising territories and space. Many would date its emergence to the Treaty of Westphalia and Romantic conceptions of "a people" who live within different territories of western Europe. Yet, despite being a recent invention, the nation state has become one of the preeminent power brokers on the world stage. The dominant actors in climate negotiations, nation states are key players within environmental governance. They also constitute the scale at which almost all statistics about the environment are produced.

Environmental historians have been particularly influential in unearthing the problematic assumptions underlying the nation state's use as a pregiven unit for research. In the first half of the twentieth century, prominent French historians presented a new way of thinking about the craft of writing history. Fernand Braudel was one of the leading scholars of the *Annales* School, as they were known, and he noted that studies of the Mediterranean long had focused on the European part of the region, and not least national political histories. But Braudel (1972) argued that the Mediterranean had to be understood as an integrated if diverse historical-geographical

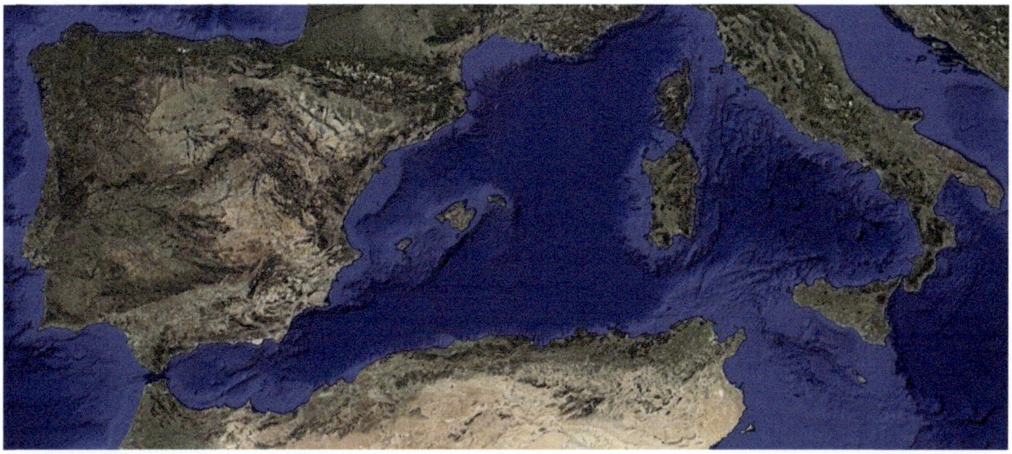

FIGURE 3.2 Where should a study of the Mediterranean begin and end?

Source: TerraMetrics/Google Maps 2022

unit that exceeded any one effort to bound it (Figure 3.2). In his preface to the first edition of *La Méditerranée*, he wrote:

> But how could one write any history of the sea, even over a period of only fifty years, if one stopped at one end with the Pillars of Hercules and at the other with the straits at whose entrance ancient Ilium once stood guard? The question of boundaries is the first to be encountered; from it all others flow. To draw a boundary around anything is to define, analyse, and reconstruct it, in this case select, indeed adopt, a philosophy of history.
>
> (18)

While complicating the methodological choices of how to define the unit of study, Braudel also problematised the concept of time, or perhaps temporal rhythms, with which the historian worked. Having lived through the violent political ruptures of the world wars, Braudel argued that history should not merely be written with a focus on individuals and events playing out over the course of a few years or decades (*histoire événementielle*). Instead, the historian had to pay careful attention to contexts, processes, mentalities, and "totalities" that far exceeded the political everyday. It is fascinating to see how Braudel spends a large part of his work on the Mediterranean describing the geology and climate of the region—processes which unfolded over what *Annales* historians called the *longue durée* (long duration). From this vantage point, he starts tracing the histories of economic systems, civilisations and states, acting on a medium-paced temporal scale. It is only then that he finds it meaningful to approach the "fireflies" of political history and everyday events.

The *Annales* School quickly gained popularity outside France and not least in Latin America. In the preceding chapters, we discussed how Latin American political ecologists have developed an analysis of modernity/coloniality, suggesting that the global economy is structured along unequal lines of class and race. The origins of this analysis can be traced back to the work of the *dependista* scholars working in the 1960s and 1970s whose work shares many of

Braudel's concerns, albeit in a much more politicised context. The *dependistas* (or "dependency theorists") argued that the economic development of the Latin American nation states could not be understood on the same terms as the development of their former colonisers. Rather than building universal economic models around "national development", it was necessary to study how each nation state was part of a larger international political economy. The capitalist world economy, as Andre Gunder Frank (1967) famously argued, was developing in a geographically contradictory fashion so that nation states in an industrial core were able to sustain their development by drawing on labour and resources from a global periphery, perpetuating the underdevelopment of the latter.

While challenging the tendency to fetishise the national scale among economists, some *dependista* scholars argued that in highlighting the relations between the national and the international levels, their analyses neglected the urban scale. For Aníbal Quijano (1975), the patterns of urbanisation in Latin America also had to be understood in relation to the Latin American countries' role as extractive economies in the world system. Again questioning the role of the nation state, he wrote: "although the centers of power that hold sway over the dependent societies are geographically outside them, and, in that sense, are external, their interdependence moulds and determines existing structures and events within the dependent societies" (124). Nevertheless, a succeeding generation of political ecologists have tempered the structural rigour of such "world systems perspectives", noting that they tend to confer explanatory power almost exclusively to global forces and "depict local peoples as having less agency than they really have" (Dove and Kammen 2015: 69).

In a Western context, the discussions around the perils of methodological nationalism find echoes in Swyngedouw's work on "glocalisation". Another way to phrase Swyngedouw's critique is that he, like Marx, emphasises processes over "things". Nation states, cities, households and bodies are constituted out of distinct sets of socio-ecological processes. If we want to understand these distinct scales, we need to get to grips with the processes that give rise to them. Take Piccadilly Circus in London as an example where Erik Swyngedouw and Maria Kaïka (2003: 568) observe the following socio-ecological processes and relations coming together:

> The coffee I sip connects me to the conditions of peasants in Colombia or Tanzania and to the Thames River Basin as much as to the climates and plants, pesticides and technologies, traders and merchants, shippers and bankers, bosses and workers. The cars burning fuels from oil-deposits and pumping CO_2 into the air, affecting forests and climates around the globe, further complete the global geographic mappings and traces that flow through the urban and "produce" London's cityscape as a palimpsest of densely layered bodily, local, national, and global—but geographically depressingly uneven—socioecological processes.

In this account, Piccadilly Circus is less a bounded place or a "thing" and more an ensemble of flows and relations.

Going even further in their pursuit of processes over things, Hillary Angelo and David Wachsmuth (2015) argue that if we are really interested in the extended relationships that constitute urban experiences, then researchers should not focus so steadfastly on individual cities. While there is an epistemological need to draw boundaries around processes if one is to map and quantify them (Guibrunet et al. 2017), scale quickly becomes a stumbling block according to Angelo and Wachsmuth (2015). Urban political ecologists have, therefore, fallen into a "methodological cityism", focusing on one scale to the neglect of the processes and relations constituting that scale. These are processes that imbue "things", such as the city and its infrastructures, with mystical powers at a given historical and geographical moment (Chapter 5).

SCALE AS STRUGGLE AND PROCESS

Throughout the discussions above, the concept of scale has been treated mainly as an epistemological phenomenon. The appropriateness of different scales—the city, the nation state or the "site"—is a topic of contestation: questions are posed around where a study should be located and at what scale processes should gain explanatory power over other ones. Nevertheless, for many environmental activists—as indeed for the social movements they work with and alongside—questions around scale are not simply about intellectual debate. They are about material issues such as how to access basic resources, and they have implications for political action such as how and where socio-ecological relations might be contested and transformed.

For the Marxist geographer Neil Smith, scale is just like this, a material phenomenon that is actively produced and therefore open to change. In his landmark text *Uneven Development*, Smith (2007) places particular emphasis on the role of capitalist social relations in the production of scale. Capitalism can be understood to be a mode of production that is fundamentally contradictory: when the globalised capitalist economy has developed, it has simultaneously overexploited its resource base, led to the pollution of ecological systems and the decimation of biodiversity. Thus, through its own growth, capitalism imperils life itself. Capturing these contradictions, ecological Marxists such as James O'Connor (1989) draw attention to the different ways in which capitalist production undermines the conditions upon which further production depends. Smith, in his turn, views the production of scale as intimately bound up in these contradictions and, in particular, in that between competition and cooperation.

As a historically specific mode of production, capitalism relies on competition between individual capitalists. Think about the competition between Pepsi and CocaCola, for example, or on a smaller scale, that between two local falafel vendors. Nevertheless, competition cannot result in total anarchy since this could endanger the system as a whole. Free and fair exchange—one of the defining features of capitalism—therefore depends on rules, laws and regulations in which the nation state plays an important stabilising role. Furthermore, capitalism thrives when resources can be pooled through the construction of infrastructures (what economic geographers would call "agglomeration effects") that enable things like the transportation of goods from sites of production to sites of consumption. Nation states also tend to be the main investors in large infrastructural projects, such as roads, grids, airports, and shipping terminals, or to offer private investors loan guarantees for their construction. For Smith, the nation state is a scalar manifestation of the struggle to manage the contradictions between all out competition and the need for some form of cooperation. Other scales can be thought of in similar terms, as an outcome of the contradictory tendencies and associated struggles comprising a capitalist mode of production. Together they constitute a "scalar fix".

However, within each scale—the household, the city, the nation state, the global, etc.—further contradictions begin to emerge. For example, the spatial integration of economic activities through free trade contradicts the self-imposed barriers erected by the nation state. Just as the scalar fix appears to provide a temporary resolution to one set of contradictions, it generates further contradictions. As Raymond Bryant and Sinéad Bailey (1997) show, the nation state is tasked with the role of protector of environmental resources at the same time as it is required to promote the very economic growth that might endanger those protections. These two roles sit uneasily alongside one another as different lobby groups push for greater environmental protections or the stripping away of such protections in the name of economic growth.

If scales are both produced and struggled over, they can also be an arena for bringing about change. Smith notably speaks of "scale jumping" as a political tactic. In his earlier work, Smith focused on the possibilities for the working class to "jump scales" and thereby move from being a class of people divided by the borders and xenophobia of the nation state to becoming a universal class seeking democratic change at the global scale. Smith's later work expanded from a singular focus on the working class to also consider how racialised and gendered social relations cohered at different scales. Perhaps reflecting the wider context in which he was living and working, he became far more attentive to cultural processes—for example, through his work on the Homeless Vehicle, an art intervention in New York City—and to gendered social relations, which was most clearly seen in his work with the geographer Sallie Marston (Marston and Smith 2001). In this expanded sense, scale was seen as the outcome of dynamic processes and social relations, as well as the acts of classed, racialised, and gendered social groups.

While Smith and Marston emphasised social and economic relations, biophysical processes are also enrolled in broader struggles around scale. Karl Zimmerer's (2000) work on irrigation in the Inca state and the Spanish Empire stands out here, but Swyngedouw's (1999) interpretation of how the river basins of Spain were reconfigured under successive political administrations is also a case in point (Figure 3.3). In the late nineteenth century, Spain's role as an imperial power on the global stage was declining, and nationally, Spanish politics was caught between those demanding a modernisation of the country and those conservative interests seeking to maintain the status quo. Swyngedouw shows how the politics of water—and irrigation in particular—embodied these differing perspectives on the country's future. Remarkably, to capture rainwater for new irrigation projects became a prized ability for those who wished to "regenerate" the "national spirit" after the postcolonial shock.

The modernising desires of the Spanish *regeneracionistas* (regeneracionists) required a re-engineering of Spanish hydrology and the production of a fundamentally new waterscape. As Swyngedouw argues, "The river basins would become the scale *par excellence* through which the modernizers tried to undermine or erode the powers of the more traditional provincial or national state bodies" (458). A river basin refers to the area of land from which water drains into a river and its tributaries. The cost of transporting water between river basins is often prohibitively high, but for the modernisers in Spain, inter-basin transfers became the solution to the large disparities in access to water across the country. By moving water, all parts of the country could modernise. The massive dam-building programme that was to take place under General Francisco Franco's fascist regime represents the culmination of this vision—a vision in which the production of socio-ecological scale was enrolled in a broader struggle for authority within the country, seen in the control over water.

FIGURE 3.3 Dam on the River Duero, Spain

Photo by iStock.com/Gannet77

ENVIRONMENTAL JUSTICE *AT SCALE*

If both Zimmerer and Swyngedouw demonstrate that the production of socio-ecological scale is a fundamentally political question, the most appropriate scale at which environmental movements may exert influence has been a long-standing question. For David Harvey (1996), the historically limited influence of the environmental justice movement can be traced to its failure to achieve larger-scale ambitions. Since he made his argument, the global school strike movement, Fridays for Future, has of course emerged, starting with fifteen-year-old Greta Thunberg deciding to stage a one-person protest outside the Swedish parliament and expanding into a worldwide movement for climate justice. Framed in Smith's terms, however, the relative ineffectiveness of the environmental justice movement, centred around local struggles, can be attributed to its inability to jump scale and thus become the transformative global force Smith hoped the working class might be.

To give Harvey's critique some context, he began by thinking about a very specific struggle close to where he lived; a struggle that unfolded while he was based at the University of Oxford.

Oxford is not only home to a renowned university but also a large car plant. In the early 1990s, the Cowley plant, so called after the area in which it is located, was slated for closure and soon to result in the loss of some of the only solid, stable working-class jobs within the city. Along with other academics, Harvey joined a project to record oral histories of the plant and to work alongside union activists challenging the plant closure. However, while deeply sympathetic to the devastation of working-class livelihoods in Oxford, Harvey could not help but ponder if a job in a car plant really is an indicator of a socially and environmentally progressive society. Would one not rather reside in a world where the stark divisions between intellectual and manual labour—so evident in a city like Oxford—are ended? And is the production of cars, which is an environmentally destructive and privatised form of transportation, really something that social movements should support?

Harvey referred to the activism around the Cowley plant as a "militant particularist" struggle, which was narrowly focused on the issue of the plant closing and the loss of jobs. It was a struggle that, no more and no less, concerned working-class livelihoods in South Oxford. Harvey then shifted focus to the neighbourhood of Love Canal in upstate New York, where a working-class community had been able to challenge the local authority for building their homes and school on a toxic dump. This was a classic case of an environmental justice struggle (Bullard 1990), and crucially, it was one led by women. In spite of being dismissed by some commentators as "hysterical housewives", these women went on to challenge both these deeply gendered stereotypes and the environmental conditions in which lives were lost and harmed through toxic waste (Newman 2001). Though inspired by individuals such as the activist Lois Gibbs and the broader movement around Love Canal, Harvey nevertheless again found himself wondering whether any global ambitions for an environmentally and socially progressive society could be realised from such a "militant particularist" struggle. Justice was served in Love Canal, but then what?

Harvey was roundly critiqued for his scalar interpretation of the environmental justice movement. The faintly dismissive tone with which he questioned the struggles in Cowley and upstate New York annoyed many environmental activists. More importantly for some, he failed to articulate one of the key aspects of his own broader approach to understanding the world: the particular—as Harvey framed it—can never be isolated from the broader processes out of which it is constituted and out of which it becomes a crucial moment. The struggles around car manufacturing or the location of toxic waste facilities always already embody processes that connect them with other sites and processes that make up the global. As Stefan Kipfer (2022) demonstrates in an important paper on the philosopher Frantz Fanon, political struggles themselves suggest a comparative approach that inevitably shapes interpretations of scale. The recent development of the environmental justice movement into a global movement for climate justice points to the ways in which Harvey missed an opportunity to analyse the larger context in which his militant particularism took place. Environmental movements can jump scale.

To be fair, Harvey's critics have often attacked something of a straw person. The former's unease around the apparent limitations of a place-based politics was much more to do with exploring what militancy could become rather than an outright dismissal of grounded struggles. Indeed, Harvey's point was that place *does* matter but that it is often not enough. Nevertheless, the most jarring, if to our eyes misplaced, critique resulted in a paper calling for "a human geography without scale". According to Marston et al. (2005), Marxists such as Harvey, Smith and Swyngedouw had become so obsessed with the production of scale and scale jumping that they were reinforcing the very scalar hierarchies they wished to critique. Instead, Marston et al.

argued that human geographers should develop a "flat ontology" that begins from the site and explores horizontal and "rhizomatic" relations (that is "root-like" and "branching", following the work of Gilles Deleuze) rather than the vertical relations so associated with the analysis of scale in political ecology. Research in this direction has recently generated some fascinating insights (see notably Rocheleau 2015 on a rhizomatic approach to political ecology), but in adopting such a perspective, we risk losing an important conceptual tool for engaging in political ecology as praxis.

Somehow the scale debate turned into a debate ever more detached from the actual struggles lived and engaged in by everyday people—people challenging extractivism, the legacies of colonial encounters and the quotidian effects of climate change. The scalar levels defined by the framework became the target of contention rather than the processes giving rise to them (Vayda and Walters 1999). Although adopting "a political ecology without scale" for us would be to go too far, impoverishing the field, it is worth reiterating the points we made earlier in this chapter as a partial response to such claims:

1 Socio-ecological processes do converge around particular patterns at particular times, and the decisions researchers make within their analyses influence whether they can (or whether they wish to) identify such patterns
2 The researchers' choices set a process of abstraction in motion, which determines whether some patterns are deemed worthy of further observation and whether these should play a role within their explanatory models or not

Wishing away the production of scale and the epistemological choices it calls for is potentially harmful to actually existing struggles. Take the example of economically marginalised populations in the global south who are exposed to the violence of climate change in the present moment. Having contributed the least to global carbon emissions, these populations now find themselves feeling the full force of global environmental change and the scaled environmental politics associated with it. To critically study such processes, and to search for political or policy-based solutions to them, require both a consideration of the material processes at work, and how they cohere within specific institutional forms at different moments.

Sadly, the scale debate began with what was an important question. In a seminal paper, Marston (2000) asked whether feminist struggles, seeking to produce greater visibility for women's work, "produced" scale. If the work of Smith and Swyngedouw demonstrated that scale is produced out of the contradictions associated with capitalist production, Marston showed how middle-class women in the United States in the early part of the twentieth century had utilised the scale of the home in order to achieve greater visibility in their struggles for equality. Katherine McKittrick (2006) also drew attention to how an auction block used for selling Black women slaves on a plantation in Virginia, more than as a site of bondage and pain, could be understood as a site of contestation through the scaled geographical relations intersecting in it. In short, scale was bound up in processes of social reproduction that included the (racialised, gendered, sexualised) body, the home, the plantation and the production of life itself.

Bringing questions of social reproduction into political ecology forces us to think of scale not only in traditionally Marxist terms of production, competition, cooperation and capital accumulation. It also requires us to think in terms of how life itself is produced and reproduced through acts of care, love, cooking, cleaning, collecting water and searching for fuelwood—acts that almost always are shaped by a deeply gendered division of labour. In a crucial moment in *Capital*, Marx finds that an understanding of capitalism that operates in the "noisy realm of exchange"—focusing on markets and trade—is severely limited. Unlocking the secrets of capitalism instead demands that we descend into the "hidden abode of production", he writes. This is the move Smith later made in his interpretation of scale, and that Matthew Huber (2022) recently has ventured in his provocative analysis of the role of the working class in climate politics. However, for Nancy Fraser (2014: 55, emphasis added), making sense of capitalism in the current context means going "*behind* Marx's hidden abode", exploring the very questions around social reproduction identified in these writings on scale. These issues remain incredibly fertile areas in which research on the political ecologies of scale might be pursued in the future.

CONCLUSION AND LEARNING OUTCOMES

Whether or not we accept the narrative that political ecology emerged as an attempt to overcome the scalar problems encountered in cultural ecology, discussions of scale have been and remain central to the field. Debates around temporal scale shape interpretations of the Anthropocene while analyses of geographical scale animate discussions around how best to both interpret and change the socio-ecological predicaments we find ourselves in. In this chapter, we have framed scale in three ways: first, it is an analytical tool that helps us distinguish among a myriad of processes, to focus on some rather than others. In this sense, scale is an *epistemological choice* or a form of abstraction that simplifies complex realities. Second, scale refers to the ways in which social and ecological processes converge in distinct patterns. Rather than a conceptual abstraction, scale is here a real abstraction with *ontological properties*. Finally, scale is a site of struggle that shapes how people might organise to transform their worlds for the better. In this sense, scale therefore relates directly to the understanding of political ecology as *praxis*.

Having read this chapter, we hope that you can:

- Distinguish between the ways in which scale is used as an analytical concept, notably in terms of temporal, observational and cartographic scales, as well as nested levels
- Explain why it is always necessary to make choices about the scales at which a study is carried out and explanation takes place
- Observe patterns of socio-ecological change in the world around you and begin to see how they cohere around specific temporal and spatial scales
- Account for the notion of "scale jumping" and its role as a political tactic for the environmental justice movement

FURTHER READING

Rangan, Haripriya, and Christian A. Kull. 2009. "What Makes Ecology 'Political'? Rethinking 'Scale' in Political Ecology." *Progress in Human Geography* 33(1): 28–45.

This article emphasises why scale is important for politicising some ecological processes over others. Rangan and Kull foreground three key moments in this process of politicisation: an ontological moment (converging patterns); an epistemological moment (analysis on the part of the researcher); and, importantly, a moment of translation through which key processes are made explicit.

Sayre, Nathan F. 2015. "Scales and Polities." In *Routledge Handbook of Political Ecology*, edited by Tom Perreault, Gavin Bridge, and James McCarthy, 504–515. London, UK: Routledge.

Sayre's chapter is a particularly lucid account of debates over scale in political ecology. In a slightly different framing from ours, he emphasises how political ecologists are interested in scale as level, size and relation. He finds the latter to be the most important.

BIBLIOGRAPHY

Angelo, Hillary, and David Wachsmuth. 2015. "Urbanizing Urban Political Ecology: A Critique of Methodological Cityism." *International Journal of Urban and Regional Research* 39(1): 16–27.

Barth, Fredrik. 2008 (1956). "Ecological Relationships of Ethnic Groups in Swat, North Pakistan." *American Anthropologist* 58: 1079–1089.

Blaikie, Piers. 1985. *The Political Economy of Soil Erosion in Developing Countries*. London, UK: Longman.

Blaikie, Piers, and Harold Brookfield, eds. 1987. *Land Degradation and Society*. London, UK: Methuen.

Braudel, Fernand. 1972. *The Mediterranean and the Mediterranean World in the Age of Philip II*, vol. 1. Berkeley, CA: University of California Press.

Bryant, Raymond L., and Sinéad Bailey. 1997. *Third World Political Ecology*. London, UK: Routledge.

Bullard, Robert D. 1990. *Dumping in Dixie: Race, Class, and Environmental Quality*. Boulder, CO: Westview Press.

Cederlöf, Gustav, and Alf Hornborg. 2021. "System Boundaries as Epistemological and Ethnographic Problems: Assessing Energy Technology and Socio-Environmental Impact." *Journal of Political Ecology* 28: 111–123.

Cheng, Wendy, and Rashad Shabazz. 2015. "Introduction: Race, Space, and Scale in the Twenty-First Century." *Occasion* 8: 1–7.

Crutzen, Paul J. 2002. "Geology of Mankind." *Nature* 415: 23.

Davis, Janae, Alex A. Moulton, Levi Van Sant, and Brian Williams. 2019. "Anthropocene, Capitalocene, … Plantationocene? A Manifesto for Ecological Justice in an Age of Global Crises." *Geography Compass* 13: e12438.

Dove, Michael R. 2006. "Equilibrium Theory and Interdisciplinary Borrowing: A Comparison of Old and New Ecological Anthropology." In *Reimagining Political Ecology*, edited by Aletta Biersack and James B. Greenberg, 43–69. Durham, NC: Duke University Press.

Dove, Michael R., and Daniel M. Kammen. 2015. *Science, Society and the Environment: Applying Anthropology and Physics to Sustainability*. London, UK: Routledge.

Edney, Matthew H. 1997. *Mapping an Empire: The Geographical Construction of British India, 1765–1843*. Chicago, IL: University of Chicago Press.

Frank, Andre Gunder. 1967. *Capitalism and Underdevelopment in Latin America*. New York, NY: Monthly Review Press.

Fraser, Nancy. 2014. "Behind Marx's Hidden Abode: For an Expanded Conception of Capitalism." *New Left Review* 86: 55–72.

Gilmore, Ruth. 2007. *Golden Gulag: Prisons, Surplus, Crisis, and Opposition in Globalizing California*. Oakland, CA: University of California Press.

Guibrunet, Louise, Martin Sanzana Calvet, and Vanesa Castán Broto. 2017. "Flows, System Boundaries and the Politics of Urban Metabolism: Waste Management in Mexico City and Santiago de Chile." *Geoforum* 85: 353–367.

Harvey, David. 1996. *Justice, Nature and the Geography of Difference*. Oxford, UK: Blackwell.

Hecht, Gabrielle. 2018. "Interscalar Vehicles for an African Anthropocene: On Waste, Temporality, and Violence." *Cultural Anthropology* 33(1): 109–114.

Hecht, Susanna B. 1985. "Environment, Development and Politics: Capital Accumulation and the Livestock Sector in Eastern Amazonia." *World Development* 13(6): 663–684.

Huber, Matthew T. 2022. *Climate Change as Class War: Building Socialism on a Warming Planet*. London, UK: Verso.

Kipfer, Stefan. 2022. "Comparison and Political Strategy: Internationalism, Colonial Rule and Urban Research after Fanon." *Urban Studies* 59(8): 1636–1654.

Kushner, Rachel. 2021. "Is Prison Necessary?" In *The Hard Crowd: Essays 2000–2020*, 175–198. New York, NY: Scribner.

Lewis, Simon L., and Mark A. Maslin. 2015. "Defining the Anthropocene." *Nature* 519(7542): 171–180.

Malm, Andreas, and Alf Hornborg. 2014. "The Geology of Mankind? A Critique of the Anthropocene Narrative." *The Anthropocene Review* 1(1): 62–69.

Marks, Robert B. 2019. *The Origins of the Modern World: A Global and Environmental Narrative from the Fifteenth to the Twenty-First Century*. Lanham, MA: Rowman & Littlefield.

Marston, Sallie A. 2000. "The Social Construction of Scale." *Progress in Human Geography* 24(2): 219–242.

Marston, Sallie A., John Paul Jones III, and Keith Woodward. 2005. "Human Geography without Scale." *Transactions of the Institute of British Geographers* 30(4): 416–432.

Marston, Sallie A., and Neil Smith. 2001. "States, Scales and Households: Limits to Scale Thinking? A Response to Brenner." *Progress in Human Geography* 25(4): 615–619.

Marx, Karl. 1976 (1867). *Capital: A Critique of Political Economy*, vol. 1. London, UK: Penguin.

McKittrick, Katherine. 2006. *Demonic Grounds: Black Women and the Cartographies of Struggle*. Minneapolis, MN: University of Minnesota Press.

Neumann, Roderick. 2005. *Making Political Ecology*. London, UK: Routledge.

Newman, Rich. 2001. "Making Environmental Politics: Women and Love Canal Activism." *Women's Studies Quarterly* 29(1): 65–84.

Newson, Linda. A. 1993. "The Demographic Collapse of Native Peoples of the Americas, 1492–1650." *Proceedings of the British Academy* 81: 247–288.

Nietschmann, Bernard. 1973. *Between Land and Water: The Subsistence Ecology of the Miskito Indians, Eastern Nicaragua*. New York, NY: Seminar Press.

O'Connor, James. 1989. "Uneven and Combined Development and Ecological Crisis: A Theoretical Introduction." *Race & Class* 30(3): 1–11.

Peluso, Nancy Lee. 1992. "The Political Ecology of Extraction and Extractive Reserves in East Kalimantan, Indonesia." *Development & Change* 23(4): 49–74.

Quijano, Aníbal. 1975. "The Urbanization of Latin American Society." In *Urbanization in Latin America: Approaches and Issues*, edited by Jorge E. Hardoy, 109–153. Garden City, NY: Anchor Books.

Rangan, Haripriya, and Christian A. Kull. 2009. "What Makes Ecology 'Political'? Rethinking 'Scale' in Political Ecology." *Progress in Human Geography* 33(1): 28–45.

Rappaport, Roy. 1968. *Pigs for the Ancestors: Ritual and Ecology of a New Guinea People*. New Haven, CT: Yale University Press.

Robbins, Paul. 2020. *Political Ecology: A Critical Introduction*, 3rd ed. Oxford, UK: Wiley-Blackwell.

Rocheleau, Dianne. 2015. "Roots, Rhizomes, Networks and Territories: Reimagining Pattern and Power in Political Ecologies." In *International Handbook of Political Ecology*, edited by Raymond L. Bryant, 70–88. Cheltenham, UK: Edward Elgar.

Rockström, Johan, et al. 2009. "A Safe Operating Space for Humanity." *Nature* 461: 472–475.

Sayre, Nathan F. 2009. "Scale." In *A Companion to Environmental Geography*, edited by Noel Castree, David Demeritt, Diana Liverman, and Bruce Rhoads, 95–108. New York, NY: John Wiley & Sons.

Sayre, Nathan F. 2015. "Scales and Polities." In *Routledge Handbook of Political Ecology*, edited by Tom Perreault, Gavin Bridge, and James McCarthy, 504–515. London, UK: Routledge.

Smith, Neil. 2007 (1984). *Uneven Development: Nature, Capital, and the Production of Space*, 3rd ed. Athens, GA: University of Georgia Press.

Steffen, Will, et al. 2015. "Planetary Boundaries: Guiding Human Development on a Changing Planet." *Science* 347(6223): 1259855.

Swyngedouw, Erik. 1999. "Modernity and Hybridity: Nature, *Regeneracionismo*, and the Production of the Spanish Waterscape, 1890–1930." *Annals of the Association of American Geographers* 89(3): 443–465.

Swyngedouw, Erik. 2004. "Globalisation or 'Glocalisation'? Networks, Territories and Rescaling." *Cambridge Review of International Affairs* 17(1): 25–48.

Swyngedouw, Erik, and Maria Kaïka. 2003. "The Environment of the City … or the Urbanization of Nature." In *A Companion to the City*, edited by Gary Bridge and Sophie Watson, 567–580. Malden, MA: Blackwell.

Vayda, Andrew P., and Bradley B. Walters. 1999. "Against Political Ecology." *Human Ecology* 27(1): 167–179.

Watts, Michael J. 2013 (1983). *Silent Violence: Food, Famine, and Peasantry in Northern Nigeria*. Athens, GA: University of Georgia Press.

Wiedmann, Thomas, Manfred Lenzen, Lorenz T. Keyβer, and Julia K. Steinberger. 2020. "Scientists' Warning on Affluence." *Nature Communications* 11(3107): 1–10.

Zimmerer, Karl S. 2000. "Rescaling Irrigation in Latin America: The Cultural Images and Political Ecology of Water Resources." *Ecumene* 7(2): 150–175.

CHAPTER 4

Nature

A love of wildlife, biodiversity and physical landscapes—alongside anger at their degradation—is what brings many researchers to political ecology, so there is something of an irony in the fact that many of us spend so much of our energy dissecting and critiquing the whole idea of nature. The basis of that critique lies partly in a non-dualistic approach to socio-environmental change: "nature" and "society" for most political ecologists are not separate spheres but mutually co-determining moments in a larger totality. However, there is also a second sense in which political ecologists are critical of nature-based claims. As Cindi Katz (1998: 56) writes, "Nature as measure and arbiter of what is good and right has had a long and powerful history, as appealing to fascists as those who would 'heal' the land". Throughout history, nature has been appropriated to give credence and permanence to political projects that are antithetical to social and environmental justice. Claims of a "natural order" can justify unequal power relations and preserve that which might be changed for the better.

Such a critical approach to nature has, unsurprisingly, been somewhat misunderstood: it has often been taken to flag a dislike for the wild, the untamed and biodiversity (Figure 4.1). Critics of the concept of nature have, therefore, been attacked by those who see themselves as nature's defenders, with the former being labelled "wilderness foes" (for example, in Dave Foreman's 1996 attack on the pathbreaking work of William Cronon 1996) or dismissed as "postmodern relativists" (Sokal 1996a). To take another example, the renowned Beat poet Gary Snyder (1996: 8) finds himself "getting a bit grumpy about the dumb arguments being put forth by high-paid intellectual types in which they are trying to knock Nature; knock the people who value Nature, and still come out smelling smart and progressive" (again, Snyder's primary target is Cronon). Another more recent writer, Paul Kingsnorth (2017: 75), takes such criticism and applies it to the environmental movement that he now feels so distanced from: "People like me, talking about birch trees and hilltops and sunsets, were politely, or less politely, elbowed to one side by people who were bringing a 'class analysis' to green politics". Nature "out there" must be defended, they argue.

Perhaps surprisingly, Kingsnorth's and Snyder's rather romantic responses to what they perceive to be an attack on nature end up mirroring the very unromantic defence of nature from the physicist Alan Sokal. Infamously, Sokal excoriated the "fashionable sectors of the American academic Left" in the 1990s, who he claimed were trying to deconstruct the laws of nature. In building his defence of nature, Sokal (1996b) submitted a paper to the respected journal *Social Text*, entitled "Toward a Transformative Hermeneutics of Quantum Gravity".

DOI: 10.4324/9781003095521-4

FIGURE 4.1 African elephant (*Loxodonta africana*) feeding off an acacia in the Serengeti

Photo by iStock.com/minarai

The paper itself "deconstructed" the law of gravity; however, when it was published, Sokal (1996a) immediately revealed it as a hoax through a follow-up piece in the journal *Lingua Franca*. This second paper set out to challenge the "epistemic relativism" that "denies the existence of objective reality", and in a rhetorical flourish, Sokal challenged "anyone who believes that the laws of physics are mere social conventions … to try transgressing those conventions from the windows of my apartment. I live on the twenty-first floor". Do not bash Nature, Snyder and Kingsnorth argue. Those who do, Sokal confirms, risk a fatal fall to earth.

We sincerely hope that we do not fall into the category of nature bashers or gravity deniers; indeed, what brings us to political ecology is a grounded concern to better understand how ecologies come to be degraded and wildlife extinct. But in this chapter, we argue that political ecologists are right to be cautious when it comes to "nature". This caution, however, need not undermine the important task of caring for nonhuman others. On the contrary, and as the quote from Katz confirms, devastating violence against both humans and nonhumans has been justified through claims to the "natural characteristics" of one group as opposed to another. Arguments about the natural relations between blood and soil—*Blut und Boden*, as it was put in Nazi Germany—have undergirded several fascist projects and are seeing a resurgence today when far-right movements are regaining political power in the midst of the climate crisis (Daggett 2018; Malm and the Zetkin Collective 2021). Unequal divisions of labour have been propped up by the claim that women have a natural propensity to perform caring duties within the confines of the home. Indeed, patriarchy, racism and heteronormativity all operate through appeals to the natural basis of gender, race and sex.

These issues highlight the fact that the concept of "nature" generates epistemological questions; questions about what discourses, narratives, beliefs and knowledges about nature *do* in relation to political and economic processes. However, while these questions are of great importance, political ecologists also think about nature in another way. There is ample evidence that the stuff we refer to as nature—or Nature for Snyder—is not a passive backdrop to human life, rather it is enrolled within political projects and transformed as part of them. If the epistemological question is how nature is constructed as an object of knowledge, and with what consequences, this sets up the ontological question of how nature is produced. This chapter delves into these debates, considering nature both as a social construction and an ontological product appearing at different scales. We start at the scale of the human body to see how these issues have been treated in ecofeminist debates and then zoom out to think about nature at the scales of populations, cities and landscapes. We conclude with some of the more recent debates taking place over conservation policies in which epistemological conceptions of nature become intertwined with new ontologies of nature.

ECOFEMINISM: "WE WON'T PLAY NATURE TO YOUR CULTURE"

One of the artist Barbara Kruger's most famous images is styled as an advertisement: it portrays a woman's face, the hair blending with the grass and the eyes covered with leaves. Typeset in black and white, a text emulating a 1980s pop magazine reads, "We won't play nature to your culture". Kruger's photographic artwork confronts the long-standing association of women with nature and men with culture. Through her creative practice, she challenges the construction of both nature and gender as well as their construction in relation to one another. Western literature, art, music and philosophy have long tended to rely on such gendered representations of nature as feminised and women as wild, uncivilised or only partly human. Thomas Gainsborough's famous painting *Mr and Mrs Andrews* (c. 1750) is just one of many to have been critiqued for these dualisms. Mrs Andrews sits at the base of a tree, immobile and rooted. In the meantime, her husband stands erect, with a shotgun in his arm, surveying the landscape. He is the Master of both the tamed landscape and his wife (Rose 1993).

The critique of women's representation—with their "leaky bodies" still supposedly rooted in "nature", as the anthropologist Emily Martin (1998) argues—is part of a much broader project involving feminist philosophers, artists and ecologists. The feminist movement emerged in the nineteenth century and focused on achieving formal emancipation for women through voting rights and legal change. In the mid-twentieth century, a second wave of feminism appeared in response to frustrations with the first. Despite their many successes, first-wave feminists had not achieved genuine equality given the continued construction of women as "other" to men. Gender, the later generation argued, was not a natural characteristic of a group that could gain political power but a social construction through which women were subjugated. Simone de Beauvoir's (2011: 608) much-quoted claim that "One is not born, but rather becomes, woman" captures this idea and became a signal phrase for efforts to "denaturalise" gendered social relations. While sex is a biological given, gender is a contingent social construction.

We will have cause to question this neat division in a moment when we turn to an understanding of the body as something "produced". For now, however, it is important to note the ways in which feminist environmentalism has been animated by tensions between those who seek to break the association of women with nature and those who see something of value in this assumed relationship. In a seminal study, Caroline Merchant (1980) identified an abrupt shift in Western thought in the seventeenth century. Through the work of Francis Bacon and others, a feminised conception of nature—personified in figures like Mother Nature or Gaia—gave way to a more instrumental interpretation of nature. Following the Scientific Revolution, nature was seen as an inert substance whose laws could be divined from the proper application of the scientific method. However, as Merchant points out, the apparent domination of nature by science was mirroring the long-standing subjugation of women through patriarchal social relations. Merchant, therefore, regards the loss of the "Earth Mother" discourse as a process that enabled the oppression of both women and nature.

Several authors have detected something of a contradiction in Merchant's position. On the one hand, she presents a feminised conception of nature as a cultural barrier to the exploitation of natural resources, but on the other, she also presents it as a trope that facilitates such exploitation. This is a tension that runs throughout ecofeminist debates more widely when different authors variously emphasise women's essential (Collard and Contrucci 1988) or non-essential (Seager 1992) association with nature. Biological ecofeminisms point to what its proponents claim are natural connections between women and nature, for example suggesting that biological reproduction and the apparent association between lunar and menstrual cycles are reasons why women have a more direct connection to nature than men. Social ecofeminisms, by contrast, challenge such claims of an innate or biological basis for women's association with nature. They instead point to the importance of socially produced divisions of labour and the social construction of gender to understand how the relationships between women and nature are established. It is important to note that these debates run through ecofeminisms developed in the global north, such as that associated with Merchant, but we could equally start from the seminal work of Wangari Maathai (2003) and the Green Belt Movement in Kenya or that of Bina Agarwal (1992) and Vandana Shiva in India (Mies and Shiva 2014) to trace the same fault lines (for a good further discussion of these different positions, see Merchant 1993).

THE NATURE OF THE BODY

If second-wave feminism's analysis of gender as a social construction was an epistemological critique, distinguishing between a natural sex and a constructed gender, it was also an ontological critique of how the embodied experiences of women under patriarchy came about. Iris Marion Young's essay "Throwing like a Girl" (1980) is a great example of how these two perspectives cross over. For Young, embodied experiences are not simply created in the minds of artists, writers and scientists, they are actually existing, lived products. Young begins her essay by considering a text from psychologist Erwin Straus, which weighs up the evidence that girls throw differently from boys. Straus (1966: 157) writes, "The girl of five does not make any use of lateral space. She does not stretch her arm sideward; she does not twist her trunk; she does not move her legs, which remain side by side". Given that such differences between "girls" and "boys" can be observed at such an early age, Straus claims they must be biological—a case of nature rather than nurture. Nevertheless, Young points out that Straus is unable to identify any physical attributes to support his explanation. Biological difference in Straus' essay appears to refer as much to some vague feminine essence as to proteins and cells.

While rejecting Straus' interpretation, Young still takes seriously the claim that girls "in advanced industrial, urban, and commercial society" (140) throw differently from boys. Against Straus' reductionist account of these differences, she tries to understand the relationship of women's bodies—and women's embodied experiences—to the societies of which they are a part. Building on de Beauvoir's attempts to situate bodies within a given set of historical, political and social circumstances, Young writes, "there is a particular style of bodily comportment which is typical of feminine existence, and this style consists of particular *modalities* of the structures and conditions of the body's existence in the world" (141, original emphasis). More than a social construction, the environment ("the structures and conditions") in which the ball is thrown influences *how* it is thrown. Patriarchy shapes how women move through space, how bodies are presented and how muscles develop. Indeed, for Young, patriarchy leaves a clear material legacy in "throwing like a girl".

Young focused explicitly on the society in which she was writing—the United States of the late 1970s—but we can transfer her analysis of how different bodies are produced to other historical moments and geographical locations. Looking at the human body in nineteenth-century Britain, for example, Sebastien Rioux (2015) examines how patriarchy influenced the diets, corporeal development and ultimately the life chances of women in this environment. In post-apartheid South Africa, where the racialised legacies of unequal water access extend into the present, women's spines are shaped by the act of carrying water and associated musculo-skeletal disorders (Geere et al. 2010). In this case, patriarchy cannot be understood as an isolated phenomenon but must be analysed through an intersectional view on power (Chapter 2): social relations of gender, race and class work together to leave a material imprint on different marked and unmarked bodies.

In recent years, the production of differently sexed bodies has been explored through the writings of Judith Butler, whose work on power and performativity we encountered in Chapter 2. Butler's arguments are complex and wide-ranging, but we can highlight a key point that stands out for our consideration of nature in this chapter. Rather than seeing gender as a social construction that is imposed on bodies, Butler views it as a performance: gender is acted out in everyday life through the medium of bodies. This performance takes place within certain constraints—the heteronormatively patterned spaces of a football stand or a riding club, for example—but performances can always be done differently. Butler's reading of gender opens up the possibility for potentially subversive, "queer" performances of it. In a central but dense statement, Butler (1993: 2) notably recasts the matter of the body "as the effect of a dynamic of power, such that the matter of bodies will be indissociable from the regulatory norms that govern their materialization and the signification of those material effects".

As a political ecologist, Julie Guthman (2015) is far more explicit about the socio-ecological relations through which different bodies come to be produced. She argues that obesity needs to be understood in relation to the political economies in which it occurs.

In a capitalist society, it is possible to profit from the sale of sugar-intensive foods but also the commodification of cures designed to resolve the problems the sale of these create. Human bodies are, of course, central to production in a capitalist economy—ingesting food and expending energy through labour—but beyond the impact of hard labour on the body, Guthman argues that capital today not only has the ability "to use the body but to *rework* bodily processes and spaces in ways that are conducive to ongoing accumulation" (2533–2534, original emphasis).

This more detailed focus on bodily metabolism is taken forward in work by Becky Mansfield and Guthman (2015). They look at recent advances in the science of epigenetics to disrupt assumptions around "nature" and "nurture". Epigenetics involves the study of how environment and behaviour can influence the ways in which genes work. The Centers for Disease Control and Prevention (2022: n.p.), which is the US public health agency, summarises the epigenetic perspective in the simplest possible terms:

> Your genes play an important role in your health, but so do your behaviors and environment, such as what you eat and how physically active you are. Epigenetics is the study of how behaviors and environment can cause changes that affect the way your genes work. Unlike genetic changes, epigenetic changes are reversible and do not change your DNA sequence, but they can change how your body reads a DNA sequence.

While much political ecological research demonstrates that social context and social practices (and therefore nurture) matter, these scientific advances seem to complicate our thinking. Epigenetic mechanisms mediate between genes and the environment, and they therefore influence the susceptibility of particular bodies to disease and changes in life expectancy. For Mansfield and Guthman, epigenetics above all provides a non-deterministic understanding of health and life chances: unlike more deterministic, marionette-like approaches, in which genes are taken to regulate life, epigenetic research highlights questions of how uneven social, political and economic opportunities act in relation to bodily processes. If Butler argued that the matter of bodies is shaped by a dynamic of power, Mansfield and Guthman assert that this dynamic has very direct biological consequences. Epigenetics shows "how oppression gets under the skin" (4).

CYBORG NATURES AND ACTOR-NETWORKS

If the body is plastic, open to different determinations, and neither fully influenced by "nature" nor "nurture", then perhaps it might be thought of as a *cyborg*. The figure of the cyborg is one of Donna Haraway's most evocative contributions to debates around nature. Haraway began her academic life in the field of zoology before completing a PhD in Biology, becoming interested in how metaphor influenced scientific experimentation. For example, why does the sperm tend to be interpreted as "brave and dynamic" but the egg as "passive and dependent" when biological reproduction is analysed (Martin 1998)?

As a feminist thinker concerned by some feminists' aversion to technology, Haraway (1991) developed the cyborg myth to help collapse the conventional dualisms between culture and nature, human and machine in Western culture. A portmanteau, the cyborg is both *cyb*ernetic and *organism*: it blurs these divisions while representing a hopeful reformulation of what might be possible in a "post-natural" world. Haraway writes:

> A cyborg is a cybernetic organism, a hybrid of machine and organism, a creature of social reality as well as a creature of fiction … The cyborg is a creature in a post-gender world … [it] is resolutely committed to partiality, irony, intimacy, and perversity. It is oppositional, utopian, and completely without innocence.
>
> (149–151)

The cyborg has had a huge influence on debates in political ecology, helping to provide a language through which a non-dualistic understanding of nature and society might be advanced. For Erik Swyngedouw (1996) and Matthew Gandy (2005), for instance, cities might be understood as cyborgs, a theme we discuss in Chapter 5.

Around the same time as Haraway wrote about the cyborg, sociologists of science began to explore the world in terms of "actor-networks" and "assemblages" of humans and nonhumans acting in unison, developing what has become known as Actor-Network Theory (ANT) (Latour 2005). In *Vibrant Matter*, Jane Bennett (2010) examines the so-called 2003 US-Canada intercontinental blackout—an electricity system failure estimated to have affected 55 million people. The blackout was not an outcome of technical or human error, she argues, but rather, electricity grids are "living, throbbing confederations" of electron flow patterns and federal legislation, nuclear reactors and economic theory, wire and lifestyles that at times of blackout are unable to cooperate (23). Instead of human agency animating the nonhuman world, a series of different "actants" (human and nonhuman) linked up in an assemblage, and agency only came about because of the temporary relationships between these different actants. "Agency", Bennett writes, "is … distributed across a mosaic" (38).

Some political ecologists have been very critical towards such a perspective. In decentring the human subject, Bennett ascribes agency to nonhumans, even investing non-living parts of ecosystems, such as a tree falling over a powerline, with an agentic "vital force". As Rebecca Lave (2015) and Alf Hornborg (2017) argue, this dilution of human intentionality and responsibility makes it very difficult to analyse questions of social power, and it appears to render emancipatory struggles rather pointless—it is not down to what humans do anyway. Gandy (2022: 33) shares this concern about a "more-than-human" political ecology that "leaves the realm of human agency in a state of ambiguity". However, curious about the possibilities of a non-anthropocentric framing of the field, he proposes an "other-than-human" approach that "retains a more nuanced sense of the scope of human subjectivities". Archie Davies (2023) instead tries to overcome the issue by drawing on the Brazilian geographer Josué de Castro and his work on the socio-biological causes of hunger. In a longer tradition of anticolonial thought, Davies develops a "metabolic humanism" that offers an embattled alternative to European humanism. These arguments bear witness to the Marxist, feminist and anticolonial roots of political ecology and the possibilities of exposing how human subjects, situated in uneven power relations, influence and are affected by environmental changes in unequal ways.

Indigenous scholars and those working closely with indigenous communities voice a different kind of criticism. Juanita Sundberg (2014) shows how more-than-human (or "posthumanist") scholarship, while critical of Western Enlightenment thought, tends to make universalising claims about the lack of attention to the nonhuman in political theory. In support of these claims, the most influential studies cite a self-referential set of Western authors and fail to recognise that the non-dualistic perspectives they advance are shared with many indigenous epistemic traditions. Just as during the Enlightenment, she argues, Western scholars are again erasing indigenous knowledges from their accounts. Reflecting on her own attempt to develop a posthumanist political ecology, Sundberg is highly self-critical:

> Why did I not seek out scholarship rooted in non-dualistic epistemic traditions? Indigenous authors in the Americas, for instance, outline complex knowledge systems wherein animals, plants, and spirits are understood as beings who participate in the everyday practices that bring worlds into being. … Does it not seem obvious to consult such work in order to think through methodological difficulties that stem from trying to understand and depict co-production from within a body of thought [i.e. Western science] that tends to purification and segmentation?
>
> (35)

For the Sami scholar Rauna Kuokkanen (2007), as Sundberg goes on to note, the lack of engagement with indigenous knowledges in the Western university perpetuates the coloniality of knowledge (Chapter 2). The systematic ignorance of non-Western voices, she argues, is part of "the ongoing exclusion of other than dominant Western epistemic and intellectual traditions" (6). In 2021, Farhana Sultana was invited as the first woman, the first person of colour, and the first person from the global south to write a "Progress report" on political ecology in the journal *Progress in Human Geography*—in its decades-long history of publishing such reports. Sultana notes in her text that many epistemologies remain at the margins of the field. Drawing on the work of bell hooks (1984), she also points out that "to be in the margin is to be aware of *both* margin and center" (Sultana 2021: 156, original emphasis), and to advance the field, it is vital to engage—though not uncritically—with knowledges and experiences relegated to the margins. These troubling criticisms notwithstanding, posthumanism, ANT and Haraway's discussion of cyborgs have confronted political ecologists with a whole new set of questions and languages, involving "imbroglios", "hybrids" and "assemblages", through which to speak about politicised environments. Interestingly, these languages often lead us back to the ecological roots of political ecology in which organisms and environments are profoundly interconnected and mutually determining moments of a larger open-ended totality.

DENATURALISING "POPULATION"

Moving from the scale of the body to the scale of population, we again find both epistemological and ontological questions at stake. For Paul Robbins (2020), the field of political ecology is partly defined by a challenge to the concept of nature advanced by the economist Thomas Malthus and those who have followed in his footsteps. The main argument in Malthus' (1798) *Essay on the Principle of Population* is well known: inevitably, population growth and the need to sustain ever-increasing numbers of bodies will outstrip resources, resulting in famine, misery, plague and war. Well-known too are some of the main criticisms of this principle, which demonstrate that population pressures can instead stimulate creativity in agricultural production (Chapter 8). Nevertheless, if these arguments are well known, the broader context in which Malthus was writing is often ignored. This context had a profound influence on Malthus' thinking on that which passed for "natural".

A sense of possibility and historical openness characterised western Europe in the wake of the French Revolution (and also other parts of the interconnected globe, such as the French colony of Saint Domingue where the slave population rebelled against its colonisers to establish the free state of Haiti). The French Revolution gave way to a short-lived but very powerful optimism among liberal philosophers, artists and writers. In the words of the poet William Wordsworth, "Bliss was it in that dawn to be alive / But to be young was very heaven!" As ever, things start to get a little complicated here for while the young Wordsworth was influenced by the sense that the natural hierarchies of the past could be changed, he was also deeply inspired by a Romantic conception of Nature as something that overwhelmed or exceeded human consciousness. In Wordsworth's poetry, Nature is elevated through lyric verse, from the heart leaping at the sight of a rainbow to the pleasure that fills his heart on the sight of a host of daffodils. For the older and more conservative Wordsworth, his anger at the outrages that the Industrial Revolution was inflicting on Nature began to cloud his sense of the possibilities for human progress. Ramachandra Guha (2000: 10–11) characterises this shift in attitude:

> England was the home of industrialization, but also of opposition to it. … In his travels Wordsworth saw only "the darker side of the great change" wrought by the Industrial Revolution: the "outrage done to nature" by the cities and factories, such that the common people were no longer "breathing fresh air" or "treading the green earth". The poet was profoundly out of sympathy with the mores of city life, with its impersonality and its elevation of money-making above all other values.

The young Wordsworth's initial response to the French Revolution, and the possibilities it seemed to present, was echoed by many others such as the radical philosopher William Godwin and the liberal thinker Nicolas de Condorcet. Interestingly, Malthus' *Essay* was addressed directly at de Condorcet. Refuting his belief in human progress, Malthus emphasised the "natural tendency" for human societies to descend into self-destructive bouts of famine and war. This tendency came down to two "laws of nature" that Malthus (1798: 4) had "discovered": first, that humans need to eat and, second, that "the passion between the sexes is necessary and will remain nearly in its present state". These were "fixed laws of our nature", Malthus reiterated, and only God could change them.

While cloaked in a language of science and nature-given circumstances, Malthus' argument was very clearly a moral one. He rejected the liberal egalitarianism associated with the French Revolution and constructed a set of heteronormative assumptions about humans in general. His favoured prescription was to target the working classes—the Irish in particular—who in Malthus' view were simply having too many children. At the time, British rule in Ireland constituted the Irish as essentially colonial subjects. Focusing on the birth rates among the poorest, it is noticeable how little concern Malthus seemed to have for the reproductive capacities of the upper classes and the rich. Pseudoscience, xenophobia, heteronormativity and a conservative opposition to the French Revolution were, therefore, cloaked in what was claimed to be "laws of nature". Given the many assumptions that shaped Malthus' *Essay*, political ecologists are perhaps justified in querying claims to "the fixed laws of our nature".

Although they were never intended to offer a perspective on biodiversity or pollution, Malthus' claims were reappropriated by vocal parts of the environmental movement in the 1960s. Population growth and Malthus' laws of human nature were now seen to be driving the destruction of external nonhuman nature. In publications such as Paul Ehrlich's *The Population Bomb* (1968) and the Club of Rome's *The Limits to Growth* (Meadows et al. 1972), a neo-Malthusian analysis developed around the "natural" relation between population pressure and environmental degradation (Chapter 8). Whereas Malthus had identified the "carnal and vicious" Irish working classes—with their love of procreation and the ale house—as the problem, the neo-Malthusians were concerned about the practices of the poorest populations of the world and argued for birth control in the global south to avert a tragedy.

David Harvey's (1974) critique of Malthus has been particularly influential in efforts to "denaturalise" these sorts of problematic assumptions. An important definition of power, as we saw in Chapter 2, is power as ideology, and for Harvey, Malthus' was an ideological worldview. The conclusions he drew served to maintain a colonial capitalist political economy in which the English upper classes could continue to draw on the resources and labour of others. Harvey argued that if critical scholars unearthed the assumptions underpinning neo-Malthusian ideology, it would be possible to adopt a radically different approach to political economy in the present. Indeed, the exposition of Malthus' *Essay* and neo-Malthusianism as ideology would fundamentally challenge the idea that capitalist social relations are somehow part of a natural order.

More recently, with grave concerns around climate change, Malthus appears to have returned with added force: is it possible to reduce carbon emissions, keeping global warming below 1.5°C, with global population numbers still rising? Even such a critical scholar as Haraway (2016: 6–7) finds herself lamenting "a 9 billion increase of human beings over 150 years, to a level of 11 billion by 2100". She goes on to argue that feminists "have not been willing to address the Great Acceleration of human numbers, fearing that to do so would be to slide once again into the muck of racism, classism, nationalism, modernism, and imperialism". While the stakes around global environmental change undoubtedly have heightened since the 1960s and 1970s, political ecologists have again mounted a critique against explanations rooted in demography: population increases are not the most significant cause of climate change.

In one forceful account, Sophie Lewis (2017) reads "Haraway against Haraway" to criticise the latter's descent into Malthusianism. Lewis foregrounds capitalism as the main driver of social inequality and environmental degradation and poses a pertinent question: "What if the cyborgs made a comeback? They knew who their enemies were. Overpopulation did not number among them. There is so much on earth, after all, that we really do have to destroy" (n.p.). In another piece, Meehan Crist (2020) develops a critique in more positive tones. For her, the defeatism of Malthusianism strips away all hope for change, but Crist gives us hope back:

> If it is human nature to burn through this planet's resources as we have for the past 150 years, then we're done; if it is human systems that do this, then we've got a chance. Modern human civilisation hasn't yet actually tried to exist sustainably on this planet: we rarely break free from the binary thinking that pits humans against the rest of nature.
>
> (n.p.)

FIGURE 4.2 K-punk mural at Goldsmiths

The quote reads: "Emancipatory politics must always destroy the appearance of a 'natural order', must reveal what is presented as necessary and inevitable to be a mere contingency, just as it must make what was previously deemed to be impossible seem attainable. Mark Fisher (1968–2017)".

Photo by Alex Loftus

To commemorate the K-punk artist and writer Mark Fisher, his students daubed the walls of Goldsmiths, University of London in the way shown in Figure 4.2. The quote from Fisher captures the essence of the understanding of a "natural order" as ideology: to "denaturalise" that which seems natural—just as Crist, Lewis and Harvey do—is to expose how an existing state of affairs comes to be rendered as fixed and immutable. However, through praxis, the world can be rendered otherwise.

CONSTRUCTING COLONIAL NATURES

Few political projects have been as successful as European colonialism in establishing the appearance of a "natural order". In *On Liberty*, one of the classic liberal treatises, John Stewart Mill (1859) brilliantly makes his case for democracy as a form of government. However, perhaps reflecting Mill's experiences as an officer in the British East India Company, he argues that just as children should not be allowed to vote, "we may leave out of consideration those backwards states of society in which the race itself may be considered as in its non-age" (23). By positioning colonised populations as biologically inferior, colonialism could be legitimised as a civilising mission: a racialised "natural order" justified profoundly unequal relations between social groups that would last for centuries. Sundberg and Jessica Dempsey (2009) argue that colonialism rested on a specific way of thinking about differences in nature. "Nature" was an area in which Europeans could define themselves as different from the colonial other. "Europeans", they write, "distinguished themselves as civilized and modern by recalling their rise above the state of nature, while those subject to violent colonization were framed as living in nature and, consequently were placed outside of modernity and civilization" (459).

The drawing of boundaries between "European" and "other" was part of the rationale for the development of Geography as an academic discipline. The French *Société de Géographie* (founded in 1821), the Berlin *Gesellschaft für Erdkunde* (1828), the Royal Geographical Society of London (1873) and the Dutch *Koninklijk Nederlandsch Aardrijkskundig Genootschap* (1873) were all professional "geographical societies" associated with exploration, mercantilism and European imperialism. Geographers could map soils and climates of tropical regions as a way of understanding the fate of the human populations living there. The "nature" of the colonies became a particularly powerful epistemological construction: as the historian David Arnold (2006) argues, India was "tropicalised" in the early nineteenth century. The diverse environments of the subcontinent were reduced to a singular abstraction of "the tropics"; a hot and humid environment conducive of uncivilised, irrational behaviour (in contrast to the temperate environments of northern Europe, of course). By the late nineteenth century, disciplinary geographers had not only established the nature of the colonies as an epistemological construct but also understood it as something with real agency, determining the fate of societies.

Alongside painters and writers, colonial geographers tended to silence indigenous populations in their accounts, writing them out of the maps and the scripts through which they charted the world (Figure 4.3). Bruce Braun (2002) examines the work of the geologist George Dawson in northern British Columbia, who in mapping this region stripped out any influence of indigenous populations. He thereby opened the way for the exploitation of this empty, "pristine" environment by colonising forces. In the 1990s, environmental protests broke out on Vancouver Island. Commercial logging interests were threatening the island's ancient temperate rainforests, the activists argued. In his study, Braun shows how the protesters thus ended up reproducing the same tropes around a "pristine" environment as Dawson did, erasing indigenous populations from their narrative.

Adding to such research, Jarrod Hore (2022) draws attention to the role of photography in settler colonialism. Photography emerged as a technology of visual representation at the same time as mythologies about settler belonging to colonised lands became entrenched in the Tasman World and the American West. Settler photography was not only an expression of environmental attachment in these different places, Hore argues, but also "a declaration of settler territoriality" (5). The landscapes depicted in the images of (formerly) aboriginal, Māori and native American lands constructed the landscape as an object of knowledge, and this construction did political work for the settlers: the settler vision of nature "was defined by the exclusion of various racial others" (6).

FIGURE 4.3 *The Founding of Australia 1788* by Algernon Talmage (1937)

This painting was commissioned to commemorate the landing of the British fleet at Sydney Cove in 1788, marking the start of British rule in Australia. Surrounded by his sailors, the central figure is Captain Arthur Phillip who led the voyage. Fully absent are the scores of convicts the British brought with them to establish a penal colony and, not least, any representatives of the Eora people on whose territories they landed.

Photo by Gustav Cederlöf

As seen in the discussion of "more-than-human" perspectives above, marginalised social groups often continue to be stripped out of many people's ways of thinking about nature, thus enhancing their marginalisation. This can be seen in the case of low-carbon biofuel production on "wastelands" in Chapter 8 (Baka 2017), but it is also evident in the area around the Little Colorado River in the United States when settler communities seek to expand their access to water resources while undermining Navajo and Hopi water rights (Curley 2021). Yet another example is the carefully packaged experiences often sold under the banner of ecotourism. Ecotourism relies on a spectacular view of "charismatic" species such as tigers and elephants—a nature sold as separate from humans. This separation is regularly made possible through the forceful appropriation of land from indigenous populations, thereby reproducing a mythical understanding of "the wild". As Asmita Kabra and Budhaditya Das (2022) show, tiger conservation in India has for a long time been based on a state-governed regime that denies resource rights to the human inhabitants of tiger reserves and displaces them through a "voluntary relocation scheme". Dan Brockington (2002) refers to this practice as "fortress conservation", which permits the spectacular experience of the ecotourist. Rather than being a space that exists apart from the human, "wilderness" is a profoundly human creation (Cronon 1996).

PRODUCING COLONIAL AND CAPITALIST NATURES

At the scale of the body, we saw how epistemological questions around nature—how "nature" is socially constructed—are often bound up in questions about how nature is produced. While the colonial encounter was and continues to be supported by a set of ideas around a distinct natural order, it has also had a profound influence on the environment as an actually existing, lived reality. In Chapter 3, we delved into discussions about the Anthropocene as "the age of the human". Human influence on the environment can be dated much further back than many of the initial accounts, which claimed the dawn of the Anthropocene as the moment in which James Watt patented the steam engine. With *Homo erectus* having first used fire around 1.5 million years ago, these timeframes could be extended considerably further. Nevertheless, as we saw in Chapter 3, for Maslin and Lewis (2015), one clearly identifiable moment of human-induced planetary change comes in the form of the "Orbis spike". This dramatic decline in CO_2 levels, which can be detected around 1610, is linked with the colonisation of the Americas and the death of millions of people through direct violence as well as the pathogens, plants and animals colonial forces brought with them. When native farmers died, trees grew back on the lands they had cultivated, leading to a measurable decline in CO_2 levels. The ensuing growth of "neo-Europe" within these regions is what the environmental historian Alfred Crosby (2004) memorably described as "ecological imperialism".

Colonialism, then, fundamentally changed nature in an ontological sense. For Anna Tsing (2015), it is impossible to look away from the experiences of slavery and dispossession that underlay this transformation of nature via processes that are both local and global in scale. She argues that the Anthropocene, therefore, is better understood in terms of the "Plantationocene". A focus on the colonial plantation economy and contemporary extractivism helps to decentre the better-known narratives of the Anthropocene that home in on the invention of the steam engine as a pivotal historical moment (Crutzen 2002). Going further still, Kathryn Yusoff (2018) questions the racial assumptions on which what she calls White Geology has constructed the Anthropocene and a particular "neutral" human subject through geological knowledge. She takes us to plantations in the Caribbean, to mines in South Africa and nuclear testing sites in the Marshall Islands to highlight the racialised violence and power relations that have characterised these places. Thus, she argues, it is necessary to acknowledge "the role of race in the production of the global spaces that constitute the Anthropocene" (66). Just like Tsing and Yusoff, Jason Moore (2015) is troubled by the anonymous human figure implied in the Anthropocene concept. For him, however, it is capitalist relations of production that should be the focal point: it is not some abstract human with a propensity to burn fossil fuels that is responsible for planetary change but a capitalist system that incentivises high-energy use and carbon emissions. "Capitalism is not an economic system", Moore writes; "it is not a social system; it is a way of organizing nature" (2). The current moment is, therefore, best known as the "Capitalocene".

Moore builds on the work of Neil Smith (2007) who, as we saw in Chapter 3, made a powerful statement about scale. Just as scale, Smith argues, nature is "produced". This idea will strike many as faintly ridiculous: one can almost hear Sokal taunting the late Smith to test his thesis from his twenty-first-floor window (and, indeed, Smith 1998 wrote a nice response to "Sokal's taunt"). For Smith, however, the idea that humans would exist separately from the "stuff" on which they depend is ridiculous. The idea that humans are only prevented from jumping out of apartment windows because science has "discovered" the law of gravity is similarly problematic. Even goats, he argues, would appear to have discovered gravity without the help of Sokal or Newton. Rather than conceiving of nature and society as separate entities, Smith tries to develop a concept of nature in which humans are not seen as separate from "it" but are agents in its ongoing production.

The "production of nature" thesis was formulated at a time when knowledge about human-induced climate change was relatively thin. Forty years later, at a time when gene editing of human organs is possible (Borg and Policante 2022), and the true scale of global environmental change is evident, Smith's claim may well seem less scandalous. Even common sense. However, as Kathleen Morrison (2018) argues, echoing aspects of Yusoff's analysis, there is a risk associated with perspectives like this—and especially the Anthropocene—that European historical experiences and chronologies are seen to be the drivers of history and environmental change. "Most proposals for an Anthropocene era", she writes, "adopt a rather limited historical perspective, assuming that significant environmental impact began only with the (European, and especially British) Industrial Revolution" (5). Rather than reproducing such a Eurocentric outlook, we should appreciate how humans in all parts of the world have shaped the matter of nature in socially unequal ways. Nevertheless, Smith's argument starts from a slightly different vantage point than most common readings of the Anthropocene or even the Capitalocene.

One possible reading of how nature is produced is to suggest that there is a "pristine nature" that has come to be radically altered under (racial) capitalism. Smith refuses such a reading and instead claims that nature *always* has been produced—throughout human history, in all parts of the world. Presumably, in such a view, the Anthropocene would have begun at least around 300,000 years ago with the evolution of modern humans in Africa. Working in a Marxist tradition, Smith understands labour to be a mediating activity in human-nature interactions, and to labour, humans use tools, machines, infrastructures and other species. We, therefore, depend on the production of our means of existence to survive. The production of those means not only involves the transformation of humans' own "nature"—as clothing, housing and cities are fashioned—but also that of the broader environments on which those same humans depend.

For Smith, capitalism is distinctive from other political economies in that it marks a radical transformation in the way nature is produced. Under capitalism, nature comes to be produced as a commodity. A commodity is a thing with use value—it has a purpose making it useful for a human—but also an exchange value: it can be traded in certain quantities for other commodities. Over time, this exchange value gains in importance and becomes definitive of the nonhuman. It levels different aspects of nature down as crude abstractions, tradable for one another. In Chapter 6, we will discuss how many indigenous communities understand water as "lifeblood", and in Chapter 8, how forest-dwelling communities see the forest as a lifeworld. Under capitalism, by contrast, water and forests are increasingly produced as abstract "resources" that can be quantified, divided up among different groups and consumed. This transformation of nature into an exchange-value abstraction is evident across a wide range of fields, from the spectacular forms of ecotourism cited above to attempts to control carbon emissions through cap-and-trade schemes (Leach and Scoones 2015). Perhaps one of the most influential areas for the commodification of nature has been the attempt to value what ecosystems provide in terms of "services" for human populations.

VALUING NATURE: SEEKING PAYMENT FOR ECOSYSTEM SERVICES

The idea of placing an economic value on the services that ecosystems provide can be traced back to the 1970s. Throughout the 1980s and 1990s, it found further backing as the academic discipline of environmental economics developed. When the United Nations launched the Millennium Ecosystem Assessment in 2001, it was clear that Payment for Ecosystem Services (PES) had gone mainstream (Adams 2020). The basic idea behind PES is simple. It suggests that ecosystems provide important services to humans. A stream, for example, can provide the service of supplying water to a brewery. Indeed, ecosystem services are often used by industries who profit from their use. Whereas the Romantic tradition called for nature to be preserved for its intrinsic qualities, the idea of ecosystem services implies that nature should be preserved for its utility for humans.

To encourage conservation, PES operates on an economistic logic through which the extent of the services ecosystems provide can and should be calculated so that the utilities and industries using them are proportionately charged. The methods for valuing nature have become increasingly sophisticated over time, even as the assessment of ecosystem services has become more extensive. PES is now one of the most important planks within global conservation efforts. Costa Rica is often cited as one of the most successful cases of a state using PES to guide its conservation practices. In 1996, Costa Rica's Forestry Law banned deforestation of private lands and initiated a programme of forest conservation based on the economic valuation of ecosystem services. The area of the country covered by forests has subsequently been estimated to have expanded from 21 per cent in 1987 to 51 per cent in 2010 (Brownson et al. 2020; see also Wallbott and Florian-Rivero 2018 for a critical account).

When nature is valued as an economic good, it at once becomes possible to trade one ecosystem for another. So-called wetland mitigation banking is an excellent example of where such policies have begun to develop. Again, the principle is simple: if a development—such as the construction of a housing unit or a shopping mall—harms a wetland, the developer can apply to restore or even create a new wetland somewhere else. If a wetland is degraded in Australia, the impact can be offset by investments in a wetland in the Philippines. In a detailed study, Morgan Robertson (2004) draws on numerous examples in the Chicago area to describe these practices as a form of neoliberal environmental governance. To enable PES, a new kind of business is created in which environmental consultancies develop the methods and standards by which it is possible to assess the ecosystem services that may be lost within a new development but also how mitigations may be undertaken elsewhere.

In Chapter 8, we explore the offsetting logic at work in the case of carbon emission trading, but a case that merits further attention here is river restoration. As with wetlands, stream restoration banking enables an individual or a company to damage a stream system in one place by investing in stream restoration in another. A large industry has been built up around this phenomenon too, with teams of consultants employed to restore nature to its supposedly pristine state. Even if the principles behind stream restoration seem scientifically rigorous, the means of assessing the extent of restoration remain poorly defined. Looking closely at actual practices, Lave and Martin Doyle (2021) argue that these assessments often come down to two basic criteria: first, the length of the river and, second, its morphology. Restored streams, therefore, tend to appear almost like an artist's impression of a riverine landscape with perfectly symmetrical meanders that bear little relation to the "imperfections" of actual rivers. As if to echo Smith's concerns about the production of an abstract nature through capitalism, Lave and Doyle show how a purely abstract stream is produced that poorly emulates the complexities of erosion and deposition in fluvial processes.

REWILDING: "NATURE KNOWS BEST"

While PES is now an established approach to conservation, a new set of ideas has recently also begun to emerge. The turn to "nature-based solutions" suggests that nature not only provides "services" but also "solutions" to environmental problems, from flooding to climate change. As in many other parts of the world, changes in land use together with the growing frequency of extreme weather events have increased the likelihood of flooding in large parts of the United Kingdom. Great hopes now rest on the shoulders of the humble beaver (Figure 4.4). This is what the charity Rewilding Britain (n.d.) writes on their website:

> The beaver is a keystone species and one of nature's most awesome ecological engineers. Through the building of dams, the digging of canals, and the creation of dead wood, beavers create and maintain habitats where an abundance and diversity of life can flourish. Dams prevent soil eroded from fields from being lost to the sea. Carbon and nutrients are trapped, improving water quality downstream. The flow of water is slowed, helping to ameliorate flooding.

Providing convincing evidence of this remarkable ability to mitigate flood events, Alan Law et al. (2017) have conducted a 12-year study on the influence of beavers on wetland restoration in Scotland. Just like Rewilding Britain, they describe the creatures as important "ecosystem engineers".

The idea of nature-based solutions draws inspiration from the "more-than-human" perspectives we touched on earlier in this chapter, rooted in Haraway's work on the cyborg and the idea of "actants" joining up in actor-networks. In a study of the controversies that followed the unauthorised release of beavers in the River Otter in the south-west of England for the purpose of rewilding, Sarah Crowley et al. (2017: 1846) find that the release can be seen as "a 'wild experiment' through which a more open-ended, experimental approach to co-inhabiting with wildlife might be attempted". Recently, Haraway (2016) herself has argued that the "solution" to climate change cannot be designated to humans alone. Instead, we must collapse the binary between human and nonhuman to "stay with the trouble" and "make-with" other species in

FIGURE 4.4 Is the beaver the solution to our environmental problems?

Linnean taxonomy identifies two species of beaver, the North American beaver (*Castor canadensis*), seen in the picture, and the Eurasian beaver (*Castor fiber*). Today, some see the beaver as a partner in conservation for its ability to maintain biodiverse ecosystems.

Photo by iStock.com/Vladone

order to create more liveable futures. True to her style, Haraway develops the bizarre concept of the "Chthulucene" and writes that "unlike either the Anthropocene or the Capitalocene, the Chthulucene is made up of ongoing multispecies stories and practices of becoming-with in times that remain at stake" (55). No longer sovereign and able to act outside their relations with others, humans must do things with other beings who, in turn, have agency. The beaver, then, can be a partner in conservation. As Haraway concludes, "We are at stake to each other" (55).

This notwithstanding, the implied shift in perspective and the practical application of nature-based solutions do not map perfectly on to each other. "Rewilding" is not a particularly new concept, but the definition of it in relation to "nature-based solutions" has recently made it gain in prominence, legitimacy and, importantly, funding (Lorimer 2015). A new generation of philanthropic capitalists (often referred to as "philanthrocapitalists") has followed in the footsteps of Douglas Tompkins—founder of the sports-clothing manufacturer North Face—and bought up large tracts to support the reintroduction of "keystone species" such as the beaver. Jeremy Leggett, whose Bunloit Rewilding is described as "a flagship for hope", has purchased vast swathes of land across Scotland. In a poignant reflection on this programme, Fraser MacDonald (2021) considers the power of landed interests in shaping the Scottish environment, analysing the dispossession of tenant farmers during the so-called Highland

Clearances in the eighteenth and nineteenth centuries as well as these more recent conservation efforts. Rewilding takes place in the context of an uneven political economy in which it plays an ideological role, he argues: "Capital drives the rewilding economy. It's the reason crofters suspect that landed power may be one part of the natural order that lairds [large estate owners] will seek to preserve" (18).

CONCLUSION AND LEARNING OUTCOMES

MacDonald's thoughtful critique takes us back to the point we set out with in this chapter. Political ecologists tend to be wary of ideas about "nature" for two key reasons. First, these ideas usually create a neat distinction between humans and the environments of which they are a part, and clearly, just as the wilderness idea is a human creation, rewilding cannot take place without humans. Second, to claim that something is "natural" is to claim that it is unaffected by human influence—that it is likely to stay that way in future and that it will follow a set of laws that only scientists have the power to really discern. However, claims of a "natural order" are often part of power-serving worldviews that legitimise unequal social, economic and political relations. If it is all natural, why should marginalised groups question the hierarchies through which, for example, women are subjugated to men or wealthy interests can buy up vast swathes of land?

Nature may well be one of the most powerful words in the English language. In this chapter, we have emphasised the epistemological politics that shape how nature is constructed as an object of knowledge. We have also discussed the ontological politics through which nature comes to be produced as a lived reality. Rather than being fixed, immutable or a constant that is separate from society, nature cannot be understood outside the cultural frameworks and the material practices that bring the human and the nonhuman together as part of a "political ecology".

Having read the chapter, we hope you can:

- Account for ecofeminist debates on the body as a social construction and a lived reality
- Interrogate epistemological frameworks that conceptualise nature and "wilderness" as distinct from society
- Explain the ideological function of claims of a "natural order"
- Analyse how historical processes, such as colonialism and capitalism, have produced nature with socially unequal outcomes
- Account for how "nature" has been used in recent conservation strategies, in work on ecosystem services and rewilding

FURTHER READING

Castree, Noel. 2005. *Nature*. London, UK: Routledge.

Castree's book demonstrates why debates over "nature" matter. Although primarily addressed at geographers—showing how their discipline has been shaped by such debates—the book is an accessible introduction for students from any discipline. It offers great further reading after this chapter.

Lave, Rebecca. 2015. "Reassembling the Structural: Political Ecology and Actor-Network Theory." In *Routledge Handbook of Political Ecology*, edited by Tom Perreault, Gavin Bridge, and James McCarthy, 213–223. London, UK: Routledge.

Lave's tightly argued chapter focuses on how political ecologists have considered the nonhuman within their research and makes a powerful case for why ANT is an *inappropriate* conceptual framework for the field. Written with both passion and clarity, it is a pleasure to read.

BIBLIOGRAPHY

Adams, W. M. 2020. *Green Development: Environment and Sustainability in a Developing World*, 4th ed. London, UK: Routledge.

Agarwal, Bina. 1992. "The Gender and Environment Debate: Lessons from India." *Feminist Studies* 18(1): 119–158.

Arnold, David. 2006. *The Tropics and the Traveling Gaze: India, Landscape, and Science, 1800–1856*. Seattle, WA: University of Washington Press.

Baka, Jennifer. 2017. "Making Space for Energy: Wasteland Development, Enclosures, and Energy Dispossessions." *Antipode: A Radical Journal of Geography* 49(4): 977–996.

Bennett, Jane. 2010. *Vibrant Matter: A Political Ecology of Things*. Durham, NC: Duke University Press.

Borg, Erica, and Amedeo Policante. 2022. *Mutant Ecologies: Manufacturing Life in the Age of Genomic Capital*. London, UK: Pluto Press.

Braun, Bruce. 2002. *The Intemperate Rainforest: Nature, Culture, and Power on Canada's West Coast*. Minneapolis, MN: University of Minnesota Press.

Brockington, Dan. 2002. *Fortress Conservation: The Preservation of the Mkomazi Game Reserve*. Melton, UK: James Currey.

Brownson, Katherine, Elizabeth P. Anderson, Susana Ferreira, Seth Wenger, Laurie Fowler, and Laura German. 2020. "Governance of Payments for Ecosystem Services Influences Social and Environmental Outcomes in Costa Rica." *Ecological Economics* 174: 106659.

Butler, Judith. 1993. *Bodies that Matter: On the Discursive Limits of "Sex"*. London, UK: Routledge.

Centers for Disease Control and Prevention. 2022. "What Is Epigenetics?" https://www.cdc.gov/genomics/disease/epigenetics.htm. Accessed: 16 September 2022.

Collard, Andrée, and Joyce Contrucci. 1988. *Rape of the Wild: Man's Violence against Animals and the Earth*. London, UK: The Women's Press.

Crist, Meehan. 2020. "Is It OK to Have a Child?" *London Review of Books* 42(5): 9–14.

Cronon, William. 1996. "The Trouble with Wilderness: Or, Getting Back to the Wrong Nature." *Environmental History* 1(1): 7–28.

Crosby, Alfred W. 2004. *Ecological Imperialism: The Biological Expansion of Europe, 900–1900*, 2nd ed. Cambridge, UK: Cambridge University Press.

Crowley, Sarah L., Steve Hinchliffe, and Robbie A. McDonald. 2017. "Nonhuman Citizens on Trial: The Ecological Politics of a Beaver Reintroduction." *Environment and Planning A: Economy and Space* 49(8): 1846–1866.

Crutzen, Paul J. 2002. "Geology of Mankind." *Nature* 415: 23.

Curley, Andrew. 2021. "Unsettling Indian Water Settlements: The Little Colorado River, the San Juan River, and Colonial Enclosures." *Antipode: A Radical Journal of Geography* 53(3): 705–723.

Daggett, Cara. 2018. "Petro-Masculinity: Fossil Fuels and Authoritarian Desire." *Millennium: Journal of International Studies* 47(1): 25–44.

Davies, Archie. 2023. *A World Without Hunger: Josué de Castro and the History of Geography*. Liverpool, UK: Liverpool University Press.

de Beauvoir, Simone. 2011 (1949). *The Second Sex*. Translated by Constance Borde and Sheila Malovany-Chevallier. New York, NY: Vintage Books.

Ehrlich, Paul R. 1968. *The Population Bomb*. New York, NY: Ballantine Books.

Foreman, Dave. 1996. "Around the Campfire." *Wild Earth* 6(4): 1–2.

Gandy, Matthew. 2005. "Cyborg Urbanization: Complexity and Monstrosity in the Contemporary City." *International Journal of Urban and Regional Research* 29(1): 26–49.

Gandy, Matthew. 2022. *Natura Urbana: Ecological Constellations in Urban Space.* Cambridge, MA: MIT Press.

Geere, Jo-Anne L., Paul R. Hunter, and Paul Jagals. 2010. "Domestic Water Carrying and Its Implications for Health: A Review and Mixed Methods Pilot Study in Limpopo Province, South Africa." *Environmental Health* 9(1): 1–13.

Guha, Ramachandra. 2000. *Environmentalism.* New Delhi, India: Oxford University Press.

Guthman, Julie. 2015. "Binging and Purging: Agrofood Capitalism and the Body as Socioecological Fix." *Environment and Planning A: Economy and Space* 47(12): 2522–2536.

Haraway, Donna J. 1991. *Simians, Cyborgs, and Women: The Reinvention of Nature.* London, UK: Free Association Books.

Haraway, Donna J. 2016. *Staying with the Trouble: Making Kin in the Chthulucene.* Durham, NC: Duke University Press.

Harvey, David. 1974. "Population, Resources, and the Ideology of Science." *Economic Geography* 50(3): 256–277.

hooks, bell. 1984. *Feminist Theory: From Margin to Center.* Boston, MA: South End Press.

Hore, Jarrod. 2022. *Visions of Nature: How Landscape Photography Shaped Settler Colonialism.* Oakland, CA: University of California Press.

Hornborg, Alf. 2017. "Artifacts Have Consequences, Not Agency: Toward a Critical Theory of Global Environmental History." *European Journal of Social Theory* 20(1): 95–110.

Kabra, Asmita, and Budhaditya Das. 2022. "Aye for the Tiger: Hegemony, Authority, and Volition in India's Regime of Dispossession for Conservation." *Oxford Development Studies* 50(1): 44–61.

Katz, Cindi. 1998. "Private Productions of Space and the 'Preservation' of Nature." In *Remaking Reality: Nature at the Millennium*, edited by Bruce Braun and Noel Castree, 45–62. London, UK: Routledge.

Kingsnorth, Paul. 2017. *Confessions of a Recovering Environmentalist.* London, UK: Faber and Faber.

Kuokkanen, Rauna. 2007. *Reshaping the University: Responsibility, Indigenous Epistemes, and the Logic of the Gift.* Vancouver, BC: UBC Press.

Latour, Bruno. 2005. *Reassembling the Social: An Introduction to Actor-Network-Theory.* Oxford, UK: Oxford University Press.

Lave, Rebecca. 2015. "Reassembling the Structural: Political Ecology and Actor-Network Theory." In *Routledge Handbook of Political Ecology*, edited by Tom Perreault, Gavin Bridge, and James McCarthy, 213–223. London, UK: Routledge.

Lave, Rebecca, and Martin Doyle. 2021. *Streams of Revenue: The Restoration Economy and the Ecosystems It Creates.* Cambridge, MA: MIT Press.

Law, Alan, Martin J. Gaywood, Kevin C. Jones, Paul Ramsay, and Nigel J. Willby. 2017. "Using Ecosystem Engineers as Tools in Habitat Restoration and Rewilding: Beaver and Wetlands." *Science of the Total Environment* 605: 1021–1030.

Leach, Melissa, and Ian Scoones, eds. 2015. *Carbon Conflicts and Forest Landscapes in Africa.* Abingdon, UK: Earthscan.

Lewis, Simon L., and Mark A. Maslin. 2015. "Defining the Anthropocene." *Nature* 519(7542): 171–180.

Lewis, Sophie. 2017. "Cthulhu Plays No Role for Me." *Viewpoint Magazine*, 8 May.

Lorimer, Jamie. 2015. *Wildlife in the Anthropocene: Conservation after Nature.* Minneapolis, MN: University of Minnesota Press.

Maathai, Wangari. 2003. *The Green Belt Movement: Sharing the Approach and the Experience.* New York, NY: Lantern Books.

MacDonald, Fraser. 2021. "Wild Beasts." *London Review of Books* 43: 18.

Malm, Andreas, and the Zetkin Collective. 2021. *White Skin, Black Fuel: On the Danger of Fossil Fascism.* London, UK: Verso.

Malthus, Thomas R. 1798. *An Essay on the Principle of Population.* London, UK: J. Johnson.

Mansfield, Becky, and Julie Guthman. 2015. "Epigenetic Life: Biological Plasticity, Abnormality, and New Configurations of Race and Reproduction." *cultural geographies* 22(1): 3–20.

Martin, Emily. 1998. "Fluid Bodies, Managed Nature." In *Remaking Reality: Nature at the Millennium*, edited by Bruce Braun and Noel Castree, 63–82. London, UK: Routledge.

Meadows, Donella H., Dennis L. Meadows, Jørgen Randers, and William W. Behrens III. 1972. *The Limits to Growth: A Report for The Club of Rome's Project on the Predicament of Mankind*. New York, NY: Universe.

Merchant, Caroline. 1980. *The Death of Nature: Women, Ecology and the Scientific Revolution*. New York, NY: Harper and Row.

Merchant, Caroline. 1993. *Radical Ecology: The Search for a Liveable World*. London, UK: Routledge.

Mies, Maria, and Vandana Shiva. 2014. *Ecofeminism*. London, UK: Zed.

Mill, John Stewart. 1859. *On Liberty*. London, UK: John W. Parker and Son.

Moore, Jason. 2015. *Capitalism in the Web of Life: Ecology and the Accumulation of Capital*. London, UK: Verso.

Morrison, Kathleen D. 2018. "Provincializing the Anthropocene: Eurocentrism in the Earth System." In *At Nature's Edge: The Global Present and Long-Term History*, edited by Gunnel Cederlöf and Mahesh Rangarajan, 1–18. New Delhi, India: Oxford University Press.

Rewilding Britain. n.d. "Reintroductions and Key Species." https://www.rewildingbritain.org.uk/explore-rewilding/reintroductions-key-species/rewilding-superstars/eurasian-beaver. Accessed: 16 September 2022.

Rioux, Sebastien. 2015. "Embodied Contradictions: Capitalism, Social Reproduction and Body Formation." *Women's Studies International Forum* 48: 194–202.

Robbins, Paul. 2020. *Political Ecology: A Critical Introduction*, 3rd ed. Oxford, UK: Wiley-Blackwell.

Robertson, Morgan M. 2004. "The Neoliberalization of Ecosystem Services: Wetland Mitigation Banking and Problems in Environmental Governance." *Geoforum* 35(3): 361–373.

Rose, Gillian. 1993. *Feminism & Geography: The Limits of Geographical Knowledge*. Minneapolis, MN: University of Minnesota Press.

Seager, Joni. 1992. *Earth Follies: Feminism, Politics and the Environment*. London, UK: Routledge.

Smith, Neil. 1998. "Afterword." In *Remaking Reality: Nature at the Millennium*, edited by Bruce Braun and Noel Castree, 269–282. London, UK: Routledge.

Smith, Neil. 2007 (1984). *Uneven Development: Nature, Capital, and the Production of Space*, 3rd ed. Athens, GA: University of Georgia Press.

Snyder, Gary. 1996. "Nature as Seen from Kitkitdizze Is No 'Social Construction'." *Wild Earth* 6(4): 7–9.

Sokal, Alan D. 1996a. "A Physicist Experiments with Cultural Studies." *Lingua Franca* 6(4): 62–64.

Sokal, Alan D. 1996b. "Transgressing the Boundaries: Toward a Transformative Hermeneutics of Quantum Gravity." *Social Text* 46/47: 217–252.

Straus, Edwin W. 1966. *Phenomenological Psychology*. New York, NY: Basic Books.

Sultana, Farhana. 2021. "Political Ecology 1: From Margins to Center." *Progress in Human Geography* 45(1): 156–465.

Sundberg, Juanita. 2014. "Decolonizing Posthumanist Geographies." *cultural geographies* 21(1): 33–47.

Sundberg, Juanita, and Jessica Dempsey. 2009. "Culture/Natures." In *International Encyclopedia of Human Geography*, edited by Rob Kitchin and Nigel Thrift, 458–463. Amsterdam, Netherlands: Springer.

Swyngedouw, Erik. 1996. "The City as a Hybrid: On Nature, Society and Cyborg Urbanization." *Capitalism Nature Socialism* 7(2): 65–80.

Tsing, Anna L. 2015. *The Mushroom at the End of the World: On the Possibility of Life in Capitalist Ruins*. Princeton, NJ: Princeton University Press.

Wallbott, Linda, and Elena M. Florian-Rivero. 2018. "Forests, Rights and Development in Costa Rica: A Political Ecology Perspective on Indigenous Peoples' Engagement in REDD+." *Conflict, Security & Development* 18(6): 493–519.

Young, Iris Marion. 1980. "Throwing Like a Girl: A Phenomenology of Feminine Body Comportment Motility and Spatiality." *Human Studies* 3(2): 137–156.

Yusoff, Kathryn. 2018. *A Billion Black Anthropocenes or None*. Minneapolis, MN: University of Minnesota Press.

CHAPTER 5
Urbanisation

When the COVID-19 pandemic forced much of the world into lockdowns, the streets and the public spaces of cities around the world appeared to empty of human life. The sound of cicadas, birds and foxes replaced the rumble of traffic, the roar of planes and the thrum of construction sites. In Uganda's capital, Kampala, the main taxi park, through which 200,000 people are said to pass on a normal day, closed. In South Africa's townships, bustling shebeens were forced to stop serving customers as curfews and a ban on the sale of alcohol came into force. Eerie ghost buses crawled the streets of London, carrying at most a handful of passengers. In response to the zoonotic virus (a virus that spreads from animals to humans, that is), the flows of people, resources and commodities that make cities so distinctive slowed or stopped (Gandy 2022a). The form and function of urban centres fundamentally altered.

Around the same time, some people hailed an "Anthropause", hoping that this period could mark a new relationship between humans and their environments. Others spoke of "nature" making a comeback, one that hubristic humanity had visited on itself. Some even spoke of nature's revenge. City life did gradually return to a form of normality albeit with some important changes. Commuter patterns changed in many cities with staggered working weeks. Some employers facilitated working-from-home arrangements. Alcohol sales in South Africa resumed although with further bans. Both airlines and public transport struggled to cope with the huge fluctuations in demand between 2020 and 2022. In Uganda, while the Kampala taxi park reopened, major new restrictions imposed on the drivers meant that few, at first, found it possible to operate.

This dramatic reconfiguration of city life draws attention to one of the most important points political ecologists make about a world that is rapidly urbanising: cities are much more than a collection of buildings, towns with cathedrals or places with populations over a certain size. They are defined by relationships, flows and practices and express ideas about how humans relate to their environments. Cities can be sources of hope or despair, just as they can be visions of the future or dystopian representations of humanity's destruction of nature. Given such a processual perspective, it is perhaps unsurprising that political ecologists often avoid simple, rigid definitions of "cityness". Some of the best research focuses not on the city as a thing but on the process of urbanisation that gives rise to the form of a city (Harvey 1996a). When analysing a place like Kampala, political ecologists are less likely to attend to the city as an architectural artefact and more likely to focus on the taxi journeys and the flows of water, energy, alcohol, waste and sewage that give rise to urban form. Indeed, to see the city as a set

DOI: 10.4324/9781003095521-5

of socio-ecological processes is to focus on the social relations, the technological mixes, the institutions and the ideas out of which "a city" appears.

In this chapter, we introduce urbanisation as a key concept in political ecology. In doing so, we place particular emphasis on the city as a distinct environment that is the product of historically and geographically specific socio-ecological processes. In what is perhaps one of the most cited maxims among urban political ecologists, this approach implies that "in a fundamental sense, there is nothing *unnatural* about New York City" (Harvey 1996b: 186, original emphasis). Just as a bird's nest, the Birds Nest Stadium in Beijing, a parking lot in Jerusalem, a hedgerow or a field system are configurations of the human and nonhuman, so on a larger scale than the stadium or the parking lot, the cities they are a part of can be understood to be the same. The challenge is to understand how specific socio-ecological processes manifest in urban environments in ways that benefit some and not others. Cities are environments in which different social groups may thrive or barely survive, just as they are habitats in which nonhuman lifeforms will flourish or die. Today, one in eight humans live in no more than 33 cities—from Tokyo to Lima—and the United Nations (2019: 1) projects that two-thirds of the world's population will be urban in 2050. At a time of a global environmental crisis driven by urban consumption, urbanisation and the enabling conditions of more just sustainable cities need to be at the centre of attention.

URBAN METABOLISMS

Just like cities, bodies are remarkable things. Bodies quietly undertake a series of chemical reactions that convert food into movement, creativity, heat and waste (Chapter 7). Drawing from the Greek word for "change", these chemical reactions—and the broader process—is referred to as metabolic. Karl Marx (1976: 283) takes the same word (albeit in German, *Stoffwechsel*) to capture the complex relationship between humans and their environments. It makes some sense to refer to human-environment relations as metabolic: the term provides a challenge to dualistic nature-society approaches and instead suggests a co-determining relationship.

It is perhaps unsurprising that many different approaches to the environment—both Marxist and more mainstream—have taken metabolism as the basis for thinking about how human-environment relations capture the processes of urbanisation. Contemplating these different uses of "metabolism", Joshua Newell and Joshua Cousins (2015) identify three overarching approaches to urban metabolism that have developed in recent years: Marxist ecologies, industrial ecology and urban ecology. They place urban *political* ecology under the broader category of "Marxist ecologies", given its explicit focus on the production and reproduction of social inequality. Thus, studies of urban metabolism have focused on the uneven outcomes resulting from flows of water (Swyngedouw 2004), food (Heynen 2006), fat (Marvin and Medd 2006), alcohol (Lawhon 2013) and energy (Luque-Ayala and Silver 2016). An important collection of essays on urban political ecology (Heynen et al. 2006) goes so far as to propose "circulations and metabolisms" as the starting point for a political ecology of the city.

A socio-ecological process, urban metabolism is one that depends on labour as its "living, form-giving fire" (Marx 1973: Notebook III/IV). You might remember Marx's understanding that humans have an active relationship with the environment from Chapter 2, transforming it through labour. To see this process in action, we might look to the remarkable rise of cities in the Gulf states. Indeed, as we write this book, the FIFA World Cup in Qatar 2022 has just been played. Already in the lead-up to it, the World Cup was an odd one: in the most blatant fashion it exposed the power relations involved in bidding for large sporting events, the lack of interest in player welfare among sports' governing bodies and, above all, the abuse of migrant workers in the Gulf states. In May 2022, NGOs including Amnesty International and Human Rights Watch addressed an open letter to Gianni Infantino, president of football's international governing body FIFA, which drew attention to the fact that migrant workers from Africa and Asia make up about 95 per cent of the Qatari workforce. They comprise almost the entire workforce behind the construction of the stadiums and the servicing of infrastructure during the World Cup. The letter goes on to note that "The scale of human rights abuses linked to the World Cup is significant. Over the last 12 years, the vast majority of migrant workers in Qatar have had to 'pay to work' in the country, with substantial illegal recruitment fees to secure their jobs, and many thousands have been subjected to widespread wage theft" (HRW 2022).

In the neighbouring United Arab Emirates, one of the most dramatic and rapid processes of urbanisation has reached its apogee in the city of Dubai. Remarkable land reclamation projects such as the Palm Jumeirah create a sense that a whole new city has emerged magically out of the sea. But as with Qatar, this apparent magic should not hide the fact that Dubai has been built through the labour of a heavily exploited migrant workforce, structured through the *kafala* employment and immigration system. As Michelle Buckley (2013: 259) writes:

> This [*kafala*] system forms a cornerstone of the autocratic state's efforts to retain power through the maintenance of a small and politically acquiescent polis, in that it has played a key role in placing very explicit limits on the state's responsibilities to non-citizen residents, while at the same time preserving the very generous entitlements—including free land, free education, subsidized energy, well-paying jobs and so on—afforded to the national citizenry.

The migrant workforce in Dubai is "the form-giving fire" within the metabolic processes giving rise to the city. Whether the dizzying heights of the Burj Khalifa or the stunning fronds—visible from space—of the Palm Jumeirah, cities rely on human labour transforming nonhuman resources such as iron, sand, saltwater and oil into useful things. Cities are an outcome of metabolic processes that bring classed, racialised and gendered social groups into an active relationship with nonhuman resources. While humans have been reliant on this metabolic exchange for subsistence throughout history, the "production of nature" has begun to take a new form under capitalism (Chapter 4). As we see in the example of Dubai, cities have increasingly come to be produced as commodities—sometimes in a highly spectacular fashion, sometimes in more mundane forms. As commodities, nature and urban space are increasingly treated as abstract, fungible things. A square metre of real estate in Dubai can be exchanged for an equivalent commodity somewhere else, and rather than valuing the qualities and the work that goes into a specific lived urban environment, emphasis is increasingly placed on the exchange value of urban space and the ability to profit from it.

MEDIATING METABOLISM THROUGH URBAN INFRASTRUCTURE

Uneven socio-ecological outcomes manifest in different ways in urban environments. One way to unpack these is to map the sprawl of urban infrastructural networks. In the South African city of Durban, for example, both water and sanitation infrastructures reflect the material legacies of apartheid with its separation of racialised groups in urban space. In Lahore, Pakistan, the levels of air pollution reflect the ways in which roads, as infrastructures for travelling, working and living, mediate urban inequalities. For several decades, environmental justice activists have drawn attention to the uneven exposure of racialised and caste-based social groups to environmental harms. As the scholar-activist Robert Bullard (1990) once demonstrated, toxic waste, air and lead pollution are all far more likely to affect communities of colour in the United States than other social groups. Moreover, in contrast to the exposure to environmental harms, a lack of access to environmental "goods", such as green space, clean water and adequate sanitation, is equally an issue of environmental justice.

While a lack of access to environmental goods is a problem that is particularly acute in some parts of the global south, it is a more significant issue than many might assume in rich countries of the world. "Plumbing poverty", as Shiloh Deitz and Katie Meehan (2019) refer to the lack of access to piped water, is an important and pernicious marker of the classed and racialised landscapes of urban and suburban United States. Plumbing poverty is a clear example of how power relations manifest in the infrastructures that lie beneath our cities. Indeed, if urban metabolisms produce uneven outcomes, these are mediated by infrastructures that often are hidden from view.

Beneath all cities lies a dense network of infrastructure without which the city would cease to function. In the Swedish city of Gothenburg, about 1,750 km of water pipes are installed below ground, bringing water to the city's residents. If the pipes, which are made from grey cast iron and polythene, were put in a straight line, they would reach as far as Rome. However, while this is a "modern" infrastructural system in a city of the global north, about 20 per cent of all drinking water in the system leaks, which can leave households without water access until utility workers can locate the damage (Malm et al. 2015). Infrastructures such as Gothenburg's and Durban's water piping are the hidden conduits through which "circulation and metabolism"—the framing concepts for a political ecology of the city (Heynen et al. 2006)—are made possible. Water networks, sewerage pipes, electricity cables, gas tubes, petrol stations, phone lines and fibre-optic cables enable businesses to run, people to heat their homes, just as they enable communication across space.

It is fascinating to have this underworld revealed when construction workers dig a hole in the street, for example during road maintenance. As we have written this book, hole after hole has been appearing in the streets of London when the utility responsible for providing water across the city, Thames Water, belatedly has been trying to plug leaks in the network. These leaks have been exacerbated by (and of course exacerbate) the drought conditions brought on by an unusually hot summer—one of the many extreme weather events linked to climate change. Large holes have punctuated the streets of Gothenburg, too, to facilitate the construction of a much-delayed train tunnel under the central city. Peering into the city's underbelly through these holes is to view a rich history of urbanisation and people's changing relationships with their broader environments. In Gothenburg, remarkably well-preserved wooden water pipes laid in the late eighteenth century were found below one street. Redundant infrastructures, made obsolescent by changes in technology, lie alongside fibreoptic cables, only recently installed to enable more rapid communications. If these infrastructural systems represent different moments in a process of urbanisation, they also speak of different historical visions of what the city might *be* (Figure 5.1).

FIGURE 5.1 Cross-section of Paris above and below ground, nineteenth century

"Sewers enjoy a special place in the pantheon of urban mythology", Matthew Gandy (1999: 24) writes of subterranean Paris. "They are one of the most intricate and multi-layered symbols and structures underlying the modern metropolis, and form a poignant point of reference for the complex labyrinth of connections that bind urban space into a coherent whole. ... Just as sewers are repeatedly associated with dirt, danger and the unseen, they are also physical manifestations of new patterns of water usage, bodily hygiene and the progressive application of new advances in science and technology".

Photo by iStock.com/Nastasic

For Maria Kaika and Erik Swyngedouw (2000: 120), technological networks "are constitutive parts of the urban … the mediators through which the perpetual process of transformation of nature into city takes place". However, these urban infrastructures—appearing as they do as solid "things"—tend to obscure the labour and the social relations involved in urban metabolic processes, such as those embarked on through the *kafala* system in Dubai. Taking the example of water, Kaika and Swyngedouw show how technological artefacts, such as dams, water towers and pumping stations, mask "the social relations through which the metabolic urbanization of water takes place" (121). While human products, technologies appear to do the work of humans (Figure 5.2). Kaika and Swyngedouw also detect something of a shift between the

FIGURE 5.2 Workers laying drainpipes in Santiago de Cuba

In a study of the construction of urban infrastructure in Cuba around 1900, we learn that the "sewer job" in Santiago de Cuba was particularly nasty business. Captain Rockenbach of the United States 12th Cavalry, who oversaw the work done by Cuban workers, stated: "'This was as difficult a piece of sewer work as could be found.' Rainfall was unusually heavy in July, water poured constantly into 1000 feet of open trench from rotten pipes until eight feet of foul water stood there and the pipes from the gas company zigzagged across the trench" (Hitchman 1975: 354).

Photo by iStock.com/ilbusca

"phantasmagoria" of an early modernity, when urban infrastructures were placed in plain view, in stunning architectural designs, to be celebrated for their remarkable achievements, and a high modernity during the twentieth century in which they have come to be buried—only visible through the holes we might occasionally be able to peer into.

While masking social relations and mediating urban metabolisms, urban infrastructures bind people together through the sharing of natural resources. Deitz and Meehan's (2019) research is an important corrective to more simplistic assumptions that all residents in cities of the global north always have access to universal service provision. Not everyone, by any means, has access to piped water and a plumbed toilet, even in a city like Gothenburg. Even so, infrastructure is often considered part of a social contract in which the state guarantees access to services in exchange for people agreeing to pay for those services, whether through taxation or user charges. This ostensible guarantee of service-provision-for-all expresses what some authors have referred to as "the modern infrastructural ideal" (Graham and Marvin 2001). An aspiration as much as an actual achievement, the modern infrastructural ideal expresses a normative vision of how cities *should* function and also how metabolic processes might produce more equal outcomes.

Nevertheless, as Steve Graham and Simon Marvin (2001) note, by the 1960s, the modern infrastructural ideal appeared to have run into trouble. This is the same period in which Matthew Gandy (2002) describes a "paranoid urbanism" beginning to develop in New York City. Coincidentally, concerns over water quality led many New Yorkers to turn to bottled water, while fiscal crises developed and investment in the maintenance of infrastructure fell to historic lows. A process of "splintering urbanism" therefore began to unfold: the modern infrastructural ideal splintered into a situation whereby some were being provided with environmental goods, and others were not. As a result, Karen Bakker (2004) describes the provision of water in many cities as occurring within "archipelagos"—interconnected islands of provision in a broader sea of unserviced locations. For urban political ecologists, the processes that produce these uneven outcomes—processes that ensure that some have access and others do not—have become the analytical focus of their work.

With their thesis of splintering urbanism, Graham and Marvin make a bold universalising claim. But urban political ecologists have criticised their suggestion that splintering also has taken place in cities of the global south for the rather easy transference of a model developed in one part of the world (the global north) to another (the global south). Indeed, Olivier Coutard (2008: 1815) concludes that "one cannot speak of 'splintering urbanism in general'—i.e., as a global trend—in any meaningful analytical way". His argument ties into a broader critique of urban studies suggesting that it must become less Eurocentric and more comparative (Roy and Ong 2011; Robinson 2016). Against the assumption that splintered provision has been a recent phenomenon in the global south, perhaps the emergence of serviced archipelagos has been the dominant trend rather than the exception for the majority world. Settler colonial contexts provide some evidence for such a claim. In South Africa, as noted, the uneven development of infrastructural networks mirrors the apartheid project's vision of "separate development" for those of "separate races". An informal settlement such as Inanda, on the outskirts of Durban, was therefore largely ignored by municipal planners until a cholera outbreak began to threaten the serviced settler-colonial population living nearby. At the same time, apartheid authorities stoked racial tensions in the area by blaming different racialised groups for the lack of adequate infrastructure (Hughes 1996). Water and sanitation networks, then, both reflected and reinforced the legacies of apartheid.

What should be clear from this discussion is that infrastructure often does more than simply provide a vessel through which resources flow. Infrastructure—just as the cities of which it is a part—is a projection of what a society would like to be. When political ecologists look at the infrastructures that lie beneath a city, they therefore tend to see much more than a black box: infrastructures perform functions that result in greater or lesser social and environmental justice. They bring a citizenry together or divide that citizenry. They ensure that people receive water for free or make sure that they pay. Crucially, the broader social and environmental relations within an urbanised society shape these technologically mediated functions, even as urban infrastructures express visions of what kind of socio-ecological relations a society might strive for in the future.

EXTENDED URBANISATION

In what is a brilliant example of this idea—that infrastructure embodies a history of ideas—Gandy (2002) charts the development of New York City's water network. He periodises the changing fortunes of municipal service provision from the emergence of a civic realm in the 1800s to the heights of "municipal urbanism" in the 1900s, followed by a gradual decline as the twentieth century progressed. "The history of cities", Gandy writes, "can be read as a history of water" (22). This urban-water history is about a gradually expanding ecological frontier that extends ever deeper into New York State, eventually covering the largest catchment area anywhere in the United States. The water network—and the socio-ecological metabolism that comprises it—expanded from the urban population's dependence on municipal wells to a network drawing from new reservoirs and storage facilities within the Croton Watershed in south-eastern New York State. As pressure came upon these supplies, the city further expanded its ecological hinterland to also include the Catskill/Delaware Watershed. The latter now accounts for 90 per cent of the water consumed within the city.

Gandy's political ecological history of New York draws attention to the fact that the boundary between city and country is a blurry one. While greenbelts, arterial roads and zip codes might suggest an outer edge, the city extends well beyond what we might think of as its physical form. Indeed, where does the city end? The urban is constituted out of relationships, circulations and metabolisms that continually stretch what we think of as a city. Focusing not on New York but Chicago, William Cronon (1992) therefore demonstrates how that city developed through its relations with a broader ecological hinterland. If the history of water cannot be separated from the history of cities (Gandy 2002), so Cronon demonstrates, Chicago's growth was intimately bound up in the growth of the grain, timber and livestock industries. Commodities markets, futures markets and a whole category of "finance" also developed within Chicago as a result of its connections to the agricultural lands of "the Great West". Thus, both Gandy's and Cronon's studies draw our attention to how the boundaries between city and country are blurred. "The city" is constituted out of its relations with other places and spaces, and while giving rise to cities, urbanisation also produces a rural hinterland supporting that urban centre.

The relational approach to urbanisation seen in both Gandy's work and in Cronon's complements the process-based understanding of urban metabolisms we discussed earlier: the city can be defined by the relations it internalises and expresses. In his writings on urbanisation, David Harvey (1973) notes that space (the space of the city, for example) can be conceptualised in three main ways. *Absolute space* is the space of private property relations—it can be divided up, measured and owned. Think of the discussion of real estate in Dubai or, perhaps closer to home, the seminar room or library in which you might be reading this book. That space is presumably owned by a university able to claim it as private property. *Relative space* is that which has often interested transport geographers and those interested in locational analysis. It is the distance between two locations as measured in length but also in time. Again, think of the distance between where you are sitting and your home, or the place you might be going to have lunch. This distance matters when thinking about where to locate a train station, a toxic dump or a hospital. *Relational space*, by contrast, captures the ways in which a city—like Chicago, New York or Kampala—is defined by its relationships with places beyond what we think of as its borders. Relational space refers to the meanings invested in a place, the practices that give it a distinct identity and the specific technologies, relations to nature and institutions that shape its use. In the case of the room in which you may or may not be sitting, relational space is what gives it meaning as an educational space, and it is what makes a lecture theatre a lecture theatre and so on.

Similar approaches have been adopted in other areas of political ecology, for example when it comes to a relational conception of an organism and its environment, human and nonhuman (Levins and Lewontin 1985). This work emphasises that in the flipside to Gandy's and Cronon's framing of New York and Chicago, the argument clearly also works the other way around: the most remote alpine meadow can be thought of as connected to—and partly constituted through—its relationship to cities. Radically extending the idea that there is no end to the city, Marcel Meili (2015) poses the following question: is the Matterhorn City? Meili's rhetoric seems perverse given that the Matterhorn usually is taken to represent everything that is the opposite of the urban. And yet, if we think of the Swiss mountain as a product of its relations with other places, spaces and environments, it has at least as many connections to cities as it does to rural parts of Switzerland (Figure 5.3). Here is Meili:

> [Matterhorn's] image stands for a mountain world that has taken over all possible urban functions, and its various simultaneous meanings have long since become entangled in conflicts that can no longer be resolved. Its image and its reality have moved close to or right into the centers. Conversely, the mountain is scarcely used or experienced as anything other than an urban monument, an athletic playground, or a nature museum.
>
> (105)

Christian Schmid (2013: 399) goes on to write that "Switzerland is today a completely urbanized country with new urban landscapes that can no longer be apprehended with the classic understanding of the city". This is a very radical claim. Meili and Schmid seem to be arguing that similar processes might be influencing Switzerland's alpine landscapes as those influencing Kampala's taxi rank. Indeed, the Atacama Desert in Northern Chile can also be thought of in similar terms, and Martín Arboleda (2020) is thereby able to trace out the many logistical networks giving rise to—and drawing from—mining interests in the desert. He shows how it

FIGURE 5.3 Is the Matterhorn city?

The Gornergrat railway station, located at 3,089 m, is a terminus for trekkers wanting to explore the area around the Matterhorn, which is visible in the background.

Photo by iStock.com/mbbirdy

becomes impossible to think of the project of extracting resources from beneath one of the driest and most barren landscapes outside relations with other port cities, other infrastructural networks and other labouring bodies around the world. The mine in the Atacama Desert is both "urban" and "planetary". Through lithium mining, it provides the resources necessary for modern battery production, which is intrinsic to smart cities, electric vehicles, mobile phone communications and the devices on which we have written this book (Chapter 7).

While Arboleda's is a deeply impressive account of the ways in which disparate ecologies are produced in relation to one another, some are more sceptical of the broader approach to "planetary urbanisation". They suggest that it represents a Eurocentric perspective, lacking analytical nuance and attention to situated processes of urbanisation (Derickson 2015). This is the same critique that the "world systems perspectives" we discussed in Chapter 3 have been exposed to. Meantime, others instead build on the thesis of planetary urbanisation to argue that urban political ecologists often have focused on the wrong things. As we also saw in Chapter 3, Hillary Angelo and David Wachsmuth (2015) detect a "methodological cityism" in studies that focus on "the city" at the expense of the relations that constitute that form. Why, they might ask, does Gandy begin his story of New York's ecological frontier in New York City and not in the Croton Watershed? Why does Swyngedouw (2004) begin his story of the Ecuadorian city of Guayaquil within the city limits? Alida Cantor (2021) discusses this point at length, and for us, her discussion points to the importance of recognising how the analytical choices we make around scale come to shape our knowledge about the world. "[A]s discussion proliferates around a planetary turn", as one of us has argued previously, "we need to keep the tension between the abstract and the concrete, the situated and the planetary, continually present within our analyses" (Loftus 2018: 93). The discussions around scale are also a helpful reminder of the need to remain focused on processes and relations rather than things. Urban political ecology, as Swyngedouw and Kaïka (2003) have it, is concerned with "the urbanization of nature" rather than "the nature of the city". Crucially, this focus on process makes it possible to envision other cities and engage in practices for a more sustainable, different kind of urbanisation.

CITIES OF THE FUTURE

In the late 1990s, a team of artists and architects, including the Harvard School of Design and the "starchitect" Rem Koolhaas, became intensely interested in the West African megalopolis of Lagos. The Nigerian capital had often been portrayed in apocalyptic tones, not least in Robert Kaplan's essay for *The Atlantic* on "The Coming Anarchy", which in neo-Malthusian terms (Chapter 4) described an African continent disintegrating under the weight of population increase, environmental degradation and political breakdown. In contrast, and inspired by a new generation of Nigerian intellectuals, Koolhaas and colleagues sensed a city brimming with possibility in Lagos. In the words of the curator and art critic Okwui Enwezor, African megacities are "centres that still hold great potential for human vitality, creativity and inventiveness" (quoted in Gandy 2005: 37–38). Although much of the work that Koolhaas and his collaborators had planned never materialised—including a major publication on the city—proposals for a new bridge project, films and various enthusiastic essays serve as records of their unexpected embrace of the city. Reflecting on this period, Koolhaas states that he was inspired by a city in which the state "had withdrawn". In spite of this withdrawal—or perhaps because of it—Koolhaas felt able to claim that Lagos "worked". The planned book was thus provisionally titled *Lagos: A City that Works.*

Although Koolhaas' rather narrow reading of what does and does not work in Lagos (as well as his slightly idiosyncratic reading of why it does) was widely criticised (e.g. Gandy 2005), other critical scholars were also seeking to learn from the cities of Africa. In doing so, they too framed them as cities of the future rather than as dystopian failures, even as they did so through careful ethnographic research and thick descriptions of the livelihood struggles out of which African urbanisation was constituted. This is clearly evident in AbdouMaliq Simone's (2004) work, where he echoes some of the language used by Koolhaas. Of the African city, he writes:

> a wide range of provisional, highly fluid, yet coordinated and collective actions are being generated that run parallel to, yet intersect with, a growing proliferation of decentralized local authorities, small-scale enterprises, community associations, and civil society organizations. These actions are, in turn, replete with locally generated moral and social economies, compelled, nevertheless, from a more expansive engagement with a broad range of external processes and actors. If African cities do, at some level, work, then I contend that these practices play a major role in making them work.
>
> (13)

As a whole, Simone's book *For the City yet to Come* is a plea to better understand informal practices and collectives—to analyse their moral and social economies and to situate these within broader histories. Thus, it also starts pointing to what the city of the future might look like.

Resonating with Simone's argument, Ananya Roy (2011) upends conventional thinking about megacities, which are cities with more than ten million inhabitants. In Mumbai, approximately 60 per cent of the population live in slums, and these areas, she notes, are often portrayed in dystopian, apocalyptic terms, as urban anomalies that need to be erased. Roy instead calls on scholars and policymakers to see the city from the perspective of the slum, without whose inhabitants' labour the city would cease to function. If the slum is seen as "a terrain of habitation, livelihood and politics" (224), how might the experiences of slum dwellers be the starting point for a more just city? On the streets of Delhi, people in cars and cycle rickshaws compete for space. Rickshaws are not only an emissions-free mode of transport and are used more frequently by women and children than by men but also provide an income to some of the most marginalised people in the city. While India's urban middle-classes often voice demands for environmental protection in rural areas, Amita Baviskar (2020: 131) notes that cars "are the big blind spot in the mirrors of [these] bourgeois environmentalists". In spite of being one of the main contributing factors to Delhi's toxic levels of air pollution, cars are seen as an unquestionable part of the urban landscape by a middle class that has sought to banish the cycle rickshaw. "Despite its obvious utility—economic, social and environmental", Baviskar writes, "the cycle-rickshaw is seen as an embarrassment in a world-class city in the making. It does not fit into the modernist visions of judges, the bourgeoisie, the Delhi government, even though its proliferation attests to its eminent functionality" (130). So, how would an environmentalism of the migrant men who pull rickshaws, or the women and children who predominantly use them, transform the city?

In a different context, Carl Death (2022) has recently found inspiration in the work of the AfricanFuturist writer Nnedi Okorafor to destabilise mainstream imaginaries of the future. Okorafor's novel *Who Fears Death* (2010), he argues, can help to "imagine new forms of being and becoming in the context of climate change" (240). Just like existing African and Asian cities, AfricanFuturist imaginaries can, therefore, provide lessons for the socio-ecological worlds we might seek to construct. Such interest in "the city yet to come" has—of course—shaped planning debates for well over a century. Contributions to these debates have almost always presented visions of the kinds of relations that need to be established between the city and its environment (this is not least evident from Koolhaas' engagement with the deltaic city of Lagos). Crucially, visions of future cities are as much socio-ecological imaginings as they are urban ones. It is interesting to note, therefore, that these socio-ecological visions historically have tended to call for a new "balance" between the city and the countryside. One of the most important texts in the field of urban design is Ebenezer Howard's (1898) *Garden Cities of Tomorrow*. Howard was appalled by the living conditions of the working classes in Britain, who often lived in slums where rivers served both as sewers and sources of drinking water—an infrastructural solution that contributed to a series of cholera pandemics during the nineteenth century. Howard instead presented a vision of a future city in which humans would live in harmony with the environment, finding a balance between the urban and the rural. His book presents an image of "three magnets" that seeks to capture the advantages and disadvantages of rural and urban life alike,

FIGURE 5.4 *Örgryte trädgårdsstad* (Örgryte Garden City), Gothenburg, Sweden

The Garden City movement spread far outside Howard's native England. It became a dominant urban planning ideal in Sweden in the 1910s and 1920s when *Örgryte trädgårdsstad* was built. In contrast to Howard's ideal, Örgryte and other Swedish "garden cities" were built on a neighbourhood scale in already existing cities, and they were not designed to be self-sufficient in the manner Howard had intended. While Howard envisioned the working class to be living in "the Garden City of tomorrow", Örgryte is today one of the areas with the highest real estate values in Gothenburg.

Photo by Gustav Cederlöf

and he then goes on to portray the many advantages of the "Town-country"—his own vision of the Garden City (Figure 5.4). Howard writes:

> I will undertake, then, to show how in "Town-country" equal, nay better, opportunities of social intercourse may be enjoyed than are enjoyed in any crowded city, while yet the beauties of nature may encompass and enfold each dweller therein; how higher wages are compatible with reduced rents and rates; how abundant opportunities for employment and bright prospects of advancement may be secured for all; how capital may be attracted and wealth created; how the most admirable sanitary conditions may be ensured; how beautiful homes and gardens may be seen on every hand; how the bounds of freedom may be widened, and yet all the best results of concert and co-operation gathered in by a happy people. (19)

While well-intentioned and no doubt progressive, Howard's vision of the Garden City—and the relations with nature that might be established in it—rest on longstanding ideological conceptions of "nature". We encountered these ideas, which rest on a fundamental separation of nature and society, in Chapter 4. The ideological approach to nature can also be seen in the writings and designs of other influential urban planners of the twentieth century (see Swyngedouw and Kaïka 2003). It is present in the designs of Patrick Geddes and Frederick Law Olmsted, the urban planner and landscape architect who co-designed Central Park in New York. It can also be seen in the work of the influential architect Frank Lloyd Wright, whose design for the Guggenheim Museum sits on the edge of Central Park, as well as in that of perhaps the most influential urban planner of the twentieth century, Le Corbusier. He was one of the key architects of the modernist movement, seeing standardisation and industrial production as the enabling forces of a modern, scientifically designed city.

Intriguingly, beyond these past visions of the city and urban design (all, interestingly, proposed by white men), Swyngedouw and Kaïka (2003) suggest that similarly dualistic interpretations of nature and society have influenced ideas around "the sustainable city". While some of the language used might be different, contemporary plans to build hugely ambitious new ecocities, such as Masdar in the United Arab Emirates, rest on a set of interpretations akin to Howard's and Le Corbusier's. Swyngedouw and Kaïka's response to such an ideological conception is to instead articulate a vision of the city as constituted out of processes that blur the boundaries between the social and the ecological, the city and the countryside, understanding these as inseparable dimensions of the same process. "[V]iewing the city as a process of continuous, but contested, socioecological change", they write, "which can be understood through the analysis of the circulation of socially and physically metabolized 'nature,' unlocks new arenas for thinking and acting on the city" (577).

Debates on sustainable urbanisation have recently given rise to ideas around "smart cities". Smart cities supposedly bring together new digital technologies with progressive planning and administration objectives to reduce carbon emissions and achieve new efficiencies (Kitchin et al. 2015). For example, "smart" electricity meters can monitor energy demand and production in real-time and, using price incentives, seek to change people's consumption behaviours to follow a more efficient pattern (Bulkeley et al. 2016). These new technologies, therefore, assume a new role for urban residents as "citizen sensors" (Gabrys 2014) or "digital care workers" (Burns and Andrucki 2021) responsible for achieving the new efficiencies expected of them by urban planners. "Smart" applications are often also discussed in traffic-system management, where digital sensors can produce real-time data on public transport arrivals, departures and usage, thus enabling a more efficient synchronisation of travel patterns and potentially reducing carbon emissions. However, political ecologists have approached these developments by asking the *who*-question that is so central to the field (Chapter 1): who benefits from the smart city and who loses out? How might the socio-ecological flows that result from such technological, political and economic mixes help bring about better or worse socio-ecological futures, and for whom? The *who*-question puts the political economic interests involved in "smart" technologies in focus, just as it does issues of ownership, control and the right to the city.

A RIGHT TO THE CITY?

In the late 1980s, several Brazilian cities began experimenting with what was termed "participatory budgeting". This was a process that opened up questions of ownership and control in radical ways: participatory budgeting involves residents of a municipality actively making decisions on how resources should be allocated and how budgets might be spent. Rather than electing representatives to make decisions for them, citizens were to make those decisions directly. Although it might appear somewhat unglamorous to be deciding on street lighting, the maintenance of a playground or the location of a municipal dump, the principle behind such an approach to governance was that better decisions would be made if people were able to participate directly in the process. As can be seen from these examples, many such decisions are very clearly political ecological concerns over waste, environment and the use of public space.

In the city of Porto Alegre, participatory budgeting came to be formalised within a three-stage process. Each stage represented a different scale of decision-making: neighbourhood assemblies, thematic assemblies and citywide coordinating sessions. Perhaps surprisingly, the experiments undertaken by these left-wing Brazilian municipalities were widely praised by neoliberal institutions such as the World Bank. The initial plan, however, was very much the brainchild of activists within the left-wing Workers' Party in Brazil (*Partido dos Trabalhadores*). As Raquel Rolnik (2013) demonstrates, the progressive vision behind participatory budgeting—which was later rolled out to more than 170 cities across Brazil—was a key influence on a set of policies known as the City Statute. The Workers' Party signed the statute into federal law in 2001, and this was later interpreted in terms of it legislating for "the right to the city".

Rolnik and other critical planning scholars assert that the right to the city is about more than simply saying that people should have a right to abide in a city (although this is obviously an important part of it). Crucially, the policies within the City Statute suggest that the right to the city is about asserting people's right to participate in *making* the city—an idea that goes back to the central role of labour in the urban metabolism. Participatory budgeting, then, is one relatively modest way in which this process might start. In hindsight, the City Statute embodied somewhat contradictory visions. On the one hand, it legislated for people's right to the city, while on the other, it also entailed a more market-led vision of urban entrepreneurialism, which gave way to a spectacular social and environmental horror show like the 2016 Rio Olympics. Even so, it must be viewed as one of the more radically democratic approaches to city governance.

In framing "the right to the city" as a cry for a more just city, it is crucial to acknowledge the Brazilian debates. In academic circles, however, the concept is most often associated with the French philosopher and sociologist Henri Lefebvre (1996). For Lefebvre, the city is a produced space that embodies social relations; it expresses imaginaries and divisions of labour. The particular ways in which these relations, imaginaries and divisions come together are specific to a given historical moment and a particular location. Thus, it is not difficult to see why urban political ecologists have claimed that a socio-ecological reading of the city is implied in Lefebvre's work—one in which different future visions and different metabolic processes converge in a distinct form. As Farhana Sultana and others have pointed out, the right to the city has important implications for how we might think about the distribution of resources within the city. How can urbanised socio-ecological flows be organised in more democratic ways? For Sultana and Alex Loftus (2013), therefore, it is impossible to separate the right to the city from the human right to water.

WHERE IS URBAN POLITICAL ECOLOGY?

It is evident from the above discussion that it is quite possible to conceptualise the call for the right to the city from the writings of a French philosopher just as it is by starting from the practices of urban planners and activists in Brazil. As we set out in Chapter 1, our aim in this book is to introduce political ecology by going beyond the established Anglo-American canon, at least in part, and to demonstrate that there is no one way of doing political ecology. To the contrary, the field must be situated in a wide range of contexts and be open to a range of voices. To this end, we can also learn from recent attempts in urban political ecology to theorise about human-environment relations from specific locations. While urban political ecologists have tended to focus on experiences in the global north perhaps more strongly than others, Simone's writings on African cities provide one instance of how a Eurocentric gaze can be displaced and non-Western experiences be sources of hope and potential. Others, such as Mary Lawhon et al. (2014) call for a "situated urban political ecology" and make a case for developing insights from the lived realities of cities in the global south. This argument in itself bears evidence to where most scholars in the subfield find themselves. By "situated", Lawhon et al. mean *in situ*, paying close attention to the historical and geographical specificities that make up urban environments. But they also mean "situated" as in Donna Haraway's (1991) attempt to recognise that knowledge always is located and partial—there is a no single disembodied gaze from which an objective view on reality might be gained.

For Nik Heynen (2014), who is one of the best-known contributors to urban political ecology, situated urban political ecology represents a second wave of research in the subfield. He distinguishes this wave from earlier work, including some of his own, which was largely influenced by Marxist theory. Instead, he notes, current scholarship is developing grounded, empirically rich interpretations that make it possible to theorise from the ground up. Heynen focuses particular attention on research concerning race and the city, such as that emerging from the Black Radical Tradition (Robinson 1983). Ranging from the work on racialised environmental injustices in the United States (Bullard 1990) (Chapter 3) to more recent insights on the relations between caste, racialisation and urban environmental inequalities in India and South Africa (Chari 2021; Ranganathan 2022), some of the most cutting-edge research in urban political ecology is now focused on such themes.

In an apparent echo of Heynen's identification of two waves, and in reference to discussions of "planetary urbanisation", Gandy (2022b: 34) has recently suggested that urban political ecology faces a crucial choice between a renewed political impetus or gradual marginalisation: "Perhaps the prefix 'urban' is no longer needed, since the field has come full circle, under an extended characterization of the urban, back to its empirical roots in political ecology, so that the question is one of the limits (or opportunities) for neo-Marxian contributions to environmental analysis, and not really a question of 'urban' versus 'non-urban' at all?" While we agree on the importance and vibrancy of the areas of research that Gandy and Heynen describe, we take issue with the sharp lines they draw between a first and a second wave of urban political ecology—as well as the somewhat partisan ways in which "renewed political impetus" comes to be circumscribed by one single author. We think it more fruitful to learn from the many exciting movements across the global north and south that continue to influence urban political ecology, providing nuanced analyses of the predicaments facing contemporary cities and their extended hinterlands.

CONCLUSION AND LEARNING OUTCOMES

There is a surprising level of disagreement among urban scholars about what a city actually *is*: the subject that defines the field eludes consensus over its definition. For political ecologists, therefore, "urbanisation" is a more generative term than "the city" as it draws attention to the flows, processes and relationships that comprise the urban experience. Indeed, the process (urbanisation) instead of the thing (the city) should be the subject for political ecology. In developing the subfield of urban political ecology, researchers have notably focused on flows of water, energy and food within the process of urbanisation. These flows are uneven: unsurprisingly, some get to benefit while others lose out; some profit while others make losses.

Rather like the human body, urbanisation can be conceived of in terms of a metabolism in which resources are transformed through labour. At the same time, past and present visions of what the city ought to be are concretised within infrastructural configurations. Historically, the post-Second World War settlement between labour and capital in the global north appeared to rest on universal access to resources, such as electricity and water, and a modern infrastructural ideal permitted such universal access. Nevertheless, changing political economic relationships, nationally and globally, led to an emerging business in basic resources, which contributed to the splintering of infrastructure and urbanisation. Just as the modern infrastructural ideal implied a vision of the city that ought to exist, so many political ecologists have been interested in the city yet to come. Out of what socio-ecological mixes might it emerge? Driven by a desire to challenge existing environmental injustices and create fairer, more democratic cities, many have approached this question with inspiration from struggles for the right to the city. These are struggles which enable us to think about political ecology and urban environments from several situated perspectives, from the global north or the global south, from more academic perspectives or more grounded ones.

Having read this chapter, we hope that you can:

- Interpret the city and its hinterlands as a process of urbanisation
- Explain why metabolism has become a useful concept for making sense of urbanisation
- Account for the uneven ways in which the flows of resources, such as water, energy and food, shape and are shaped by urban inequalities
- Evaluate some situated claims around cities of the future, including those inspired by visions and practices in Africa, Asia and Latin America
- Account for "the right to the city" as a radical proposal for city governance

FURTHER READING

Deitz, Shiloh, and Katie Meehan. 2019. "Plumbing Poverty: Mapping Hot Spots of Racial and Geographic Inequality in US Household Water Insecurity." *Annals of the American Association of Geographers* 109(4): 1092–1109.

In this article, Deitz and Meehan focus on the interrelationship between infrastructure, urban space and social inequality in the context of water access in US cities. Using census data, they find "plumbing poverty" to be a racialised phenomenon: the likelihood of suffering from water insecurity increases dramatically if you live in a Native American, Black or Hispanic household. It is a classic case of urban political ecological research.

Swyngedouw, Erik, and Maria Kaïka. 2003. "The Environment of the City … or the Urbanization of Nature." In *A Companion to the City*, edited by Gary Bridge and Sophie Watson, 567–580. Oxford, UK: Blackwell.

This is an accessible text that offers an excellent further introduction to key concepts in urban political ecology. Swyngedow and Kaïka argue that it is wrong to focus on "the nature of the city"—which draws the mind to the city as a "thing"—and that we instead should attend to "the urbanisation of nature" as an uneven socio-ecological process.

BIBLIOGRAPHY

Angelo, Hillary, and David Wachsmuth. 2015. "Urbanizing Urban Political Ecology: A Critique of Methodological Cityism." *International Journal of Urban and Regional Research* 39(1): 16–27.

Arboleda, Martín. 2020. *Planetary Mine: Territories of Extraction under Late Capitalism*. London, UK: Verso.

Bakker, Karen. 2004. "Archipelagos and Networks: Urbanization and Water Privatization in the South." *The Geographical Journal* 169(4): 328–341.

Baviskar, Amita. 2020. *Uncivil City: Ecology, Equity and the Commons in Delhi*. New Delhi, India: Sage.

Buckley, Michelle. 2013. "Locating Neoliberalism in Dubai: Migrant Workers and Class Struggle in the Autocratic City." *Antipode: A Radical Journal of Geography* 45(2): 256–274.

Bulkeley, Harriet, Gareth Powells, and Sandra Bell. 2016. "Smart Grids and the Constitution of Social Electricity Conduct." *Environment and Planning A: Economy and Society* 48(1): 7–23.

Bullard, Robert. 1990. *Dumping in Dixie: Race, Class, and Environmental Quality*. London, UK: Routledge.

Burns, Ryan, and Max Andrucki. 2021. "Smart Cities: Who Cares?" *Environment and Planning A: Economy and Space* 53(1): 12–30.

Cantor, Alida. 2021. "Hydrosocial Hinterlands: An Urban Political Ecology of Southern California's Hydrosocial Territory." *Environment and Planning E: Nature and Space* 4(2): 451–474.

Chari, Sharad. 2021. "The Ocean and the City: Spatial Forgeries of Racial Capitalism." *Environment and Planning D: Society and Space* 39(6): 1026–1042.

Coutard, Olivier. 2008. "Placing Splintering Urbanism: Introduction." *Geoforum* 39(6): 1815–1820.

Cronon, William. 1992. *Nature's Metropolis: Chicago and the Great West*. New York, NY: W.W. Norton & Co.

Death, Carl. 2022. "Africanfuturist Socio-Climatic Imaginaries and Nnedi Okorafor's Wild Necropolitics." *Antipode: A Radical Journal of Geography* 54(1): 240–258.

Deitz, Shiloh, and Katie Meehan. 2019. "Plumbing Poverty: Mapping Hot Spots of Racial and Geographic Inequality in US Household Water Insecurity." *Annals of the American Association of Geographers* 109(4): 1092–1109.

Derickson, Kate D. 2015. "Urban Geography 1: Locating Urban Theory in the 'Urban Age'." *Progress in Human Geography* 39(5): 647–657.

Gabrys, Jennifer. 2014. "Programming Environments: Environmentality and Citizen Sensing in the Smart City." *Environment and Planning D: Society and Space* 32(1): 30–48.

Gandy, Matthew. 1999. "The Paris Sewers and the Rationalization of Urban Space." *Transactions of the Institute of British Geographers* 24(1): 23–44.

Gandy, Matthew. 2002. *Concrete and Clay: Reworking Nature in New York City*. Cambridge, MA: MIT Press.

Gandy, Matthew. 2005. "Learning from Lagos." *New Left Review* 33: 37–52.

Gandy, Matthew. 2022a. "The Zoonotic City: Urban Political Ecology and the Pandemic Imaginary." *International Journal of Urban and Regional Research* 46(2): 202–219.

Gandy, Matthew. 2022b. "Urban Political Ecology: A Critical Reconfiguration." *Progress in Human Geography* 46(1): 21–43.

Graham, Steve, and Simon Marvin. 2001. *Splintering Urbanism: Networked Infrastructures, Technological Mobilities and the Urban Condition*. London, UK: Routledge.

Haraway, Donna. 1991. *Simians, Cyborgs, and Women: The Reinvention of Nature*. London, UK: Free Association Books.

Harvey, David. 1973. *Social Justice and the City*. London, UK: Arnold.

Harvey, David. 1996a. "Cities or Urbanisation?" *City* 1(1–2): 38–61.

Harvey, David. 1996b. *Justice, Nature and the Geography of Difference*. Oxford, UK: Blackwell.

Heynen, Nik. 2006. "Justice of Eating in the City: The Political Ecology of Urban Hunger." In *In the Nature of Cities: Urban Political Ecology and the Politics of Urban Metabolism*, edited by Nik Heynen, Maria Kaika, and Erik Swyngedouw, 144–157. London, UK: Routledge.

Heynen, Nik. 2014. "Urban Political Ecology 1: The Urban Century." *Progress in Human Geography* 38(4): 598–604.

Heynen, Nik, Maria Kaika, and Erik Swyngedouw. 2006. "Urban Political Ecology: Politicizing the Production of Urban Natures." In *In the Nature of Cities: Urban Political Ecology and the Politics of Urban Metabolism*, edited by Nik Heynen, Maria Kaika, and Erik Swyngedouw, 16–35. London, UK: Routledge.

Hitchman, James H. 1975. "Unfinished Business: Public Works in Cuba, 1898–1902." *The Americas* 31(3): 335–359.

Howard, Ebenezer. 1898. *Garden Cities of Tomorrow* (originally published as *To-morrow: A Peaceful Path to Real Reform*). London, UK: Swan Sonnenschein & Co.

HRW. 2022. "Qatar: Joint Open Letter to Gianni Infantino Regarding Remedy for Labor Abuses behind the 2022 World Cup", 17 May. https://www.hrw.org/news/2022/08/26/qatar-joint-open-letter-gianni-infantino-regarding-remedy-labor-abuses-behind-2022. Accessed 23 September 2022.

Hughes, Heather. 1996. "The City Closes in: The Incorporation of Inanda into Metropolitan Durban." In *The People's City: African Life in Twentieth-Century Durban*, edited by Paul Maylam and Iain Edwards, 299–306. Pietermaritzburg, South Africa: University of Natal Press.

Kaika, Maria, and Erik Swyngedouw. 2000. "Fetishizing the Modern City: The Phantasmagoria of Urban Technological Networks." *International Journal of Urban and Regional Research* 24(1): 120–138.

Kaplan, Robert. 1994. "The Coming Anarchy: How Scarcity, Crime, Overpopulation, Tribalism, and Disease Are Rapidly Destroying the Social Fabric of Our Planet." *The Atlantic*, February. https://www.theatlantic.com/magazine/archive/1994/02/the-coming-anarchy/304670. Accessed: 23 September 2022.

Kitchin, Rob, Tracey P. Lauriault, and Gavin McArdle. 2015. "Smart Cities and the Politics of Urban Data." In *Smart Urbanism: Utopian Vision or False Dawn*, edited by Simon Marvin, Andrés Luque-Ayala, and Colin McFarlane, 16–33. London, UK: Routledge.

Lawhon, Mary. 2013. "Flows, Friction and the Sociomaterial Metabolization of Alcohol." *Antipode: A Radical Journal of Geography* 45(3): 681–701.

Lawhon, Mary, Henrik Ernstson, and Jonathan Silver. 2014. "Provincializing Urban Political Ecology: Towards a Situated UPE through African Urbanism." *Antipode: A Radical Journal of Geography* 46(2): 497–516.

Lefebvre, Henri. 1996. *Writings on Cities*. Translated and edited by Eleonore Kofman and Elizabeth Lebas. Oxford, UK: Blackwell.

Levins, Richard, and Richard Lewontin. 1985. *The Dialectical Biologist*. Cambridge, MA: Harvard University Press.

Loftus, Alex. 2018. "Planetary Concerns." *City* 22(1): 88–95.

Luque-Ayala, Andrés, and Jonathan Silver, eds. 2016. *Energy, Power and Protest on the Urban Grid: Geographies of the Electric City*. London, UK: Routledge.

Malm, Annika, Frida Moberg, Lars Rosén, and Thomas J. R. Pettersson. 2015. "Cost-Benefit Analysis and Uncertainty Analysis of Water Loss Reduction Measures: Case Study of the Gothenburg Drinking Water Distribution System." *Water Resources Management* 29: 5451–5468.

Marvin, Simon, and Will Medd. 2006. "Metabolisms of Obe*city*: Flows of Fat through Bodies, Cities, and Sewers." *Environment and Planning A: Economy and Space* 38(2): 313–324.

Marx, Karl. 1973 (1857). *The Grundrisse*. Harmondsworth, UK: Penguin.

Marx, Karl. 1976 (1867). *Capital: A Critique of Political Economy*, vol. 1. Harmondsworth, UK: Penguin.

Meili, Marcel. 2015. "Is the Matterhorn City?" In *Implosions/Explosions: Towards a Study of Planetary Urbanization*, edited by Neil Brenner, 103–108. Berlin: JOVIS Verlag.

Newell, Joshua P., and Joshua J. Cousins. 2015. "The Boundaries of Urban Metabolism: Towards a Political-Industrial Ecology." *Progress in Human Geography* 39(6): 702–728.

Ranganathan, Malini. 2022. "Caste, Racialization, and the Making of Environmental Unfreedoms in Urban India." *Ethnic and Racial Studies* 45(2): 257–277.

Robinson, Cedric. 1983. *Black Marxism: The Making of the Black Radical Tradition*. London, UK: Zed Press.

Robinson, Jennifer. 2016. "Comparative Urbanism: New Geographies and Cultures of Theorizing the Urban." *International Journal of Urban and Regional Research* 40(1): 187–199.

Rolnik, Raquel. 2013. "Ten Years of the City Statute in Brazil: From the Struggle for Urban Reform to the World Cup Cities." *International Journal of Urban Sustainable Development* 5(1): 54–64.

Roy, Ananya. 2011. "Slumdog Cities: Rethinking Subaltern Urbanism." *International Journal of Urban and Regional Research* 35(2): 223–238.

Roy, Ananya, and Aihwa Ong, eds. 2011. *Worlding Cities: Asian Experiments and the Art of Being Global*. Oxford, UK: Wiley-Blackwell.

Schmid, Christian. 2013. "The Urbanization of Switzerland." In *Implosions/Explosions: Towards a Study of Planetary Urbanization*, edited by Neil Brenner, 90–102. Berlin: JOVIS Verlag.

Simone, AbdouMaliq. 2004. *For the City yet to Come: Changing African Life in Four Cities*. Durham, NC: Duke University Press.

Sultana, Farhana, and Alex Loftus. 2013. *The Right to Water: Politics, Governance and Social Struggles*. London, UK: Routledge.

Swyngedouw, Erik. 2004. *Social Power and the Urbanization of Water: Flows of Power*. Oxford, UK: Oxford University Press.

Swyngedouw, Erik, and Maria Kaïka. 2003. "The Environment of the City … or the Urbanization of Nature." In *A Companion to the City*, edited by Gary Bridge and Sophie Watson, 567–580. Oxford, UK: Blackwell.

United Nations. 2019. *World Urbanization Prospects 2018: Highlights*. New York, NY: United Nations.

CHAPTER 6

Water

Water is an intimate resource. "My body—like yours—primarily comprises water", the feminist scholar Astrida Neimanis (2013: 23, original emphasis) writes, and continues:

> My existence as a body of water is a biological fact, but living my embodiment as watery—embedded in a world that I share with other human, animal, vegetable, geophysical and meteorological bodies that also comprise water—has other implications as well. *We are all bodies of water.*

Building on writers such as Adrienne Rich, Neimanis explores the "biological fact" of water as a connective tissue between bodies to expand our sense of what it means to be human. Sophie Lewis (2018: 160) takes the project further, reminding us that all human life begins in water: "all humans in history have been manufactured underwater in amniotic fluid", she writes. Describing a live birth taking place at Standing Rock in North Dakota, Lewis refers to the dozens of indigenous midwives who say that "water is life", and she draws attention to the crucial work of women as "water protectors" in the struggles against fossil fuel companies at Standing Rock (Chapter 2) and the struggles of women to preserve life. The intimate relationship with water, then, extends from the bodily scale and beyond.

As each of these writers and activists make clear, water is both a potent metaphor and an important biological fact. This can make us rethink what it means to be a "more-than-human" species, but also how humans relate to one another and to the broader environments of which we are a part. Water makes up not only most of our bodies but also most of the surface of the earth. It is fundamental to all forms of life and ecosystem functions. Water also provided the energy source through which the Industrial Revolution was born (Chapter 7); oceans, seas and rivers offered the means by which slavery and colonialism were constituted. Perversely, hundreds of millions struggle on a daily basis to access water, and thousands die from drinking low-quality supplies. This "produced scarcity" attests to how unequal access arrangements shape the distribution of water locally and, at larger scales, in struggles between nation states, multinational corporations and multilateral organisations. At other times, it is too much water that is the problem, such as when a torrential monsoon in the summer of 2022 followed a prolonged heatwave, reportedly leaving one-third of Pakistan submerged (Sands 2022). Perhaps inevitably, water invites a political ecological analysis. In this chapter, we examine how political ecologists have investigated the politics of water, and in doing so,

DOI: 10.4324/9781003095521-6

we consider how a history of human societies and human-environment relations may be narrated if we start from water.

FROM THE HYDROLOGIC CYCLE TO THE HYDROSOCIAL CYCLE

The idea of the "hydrologic cycle" is so phenomenally popular that almost every school child around the world will encounter it at some point in their early years. Posters of the hydrologic cycle adorn the walls of hundreds of thousands of schools, displaying in slightly different ways how water circulates from the ground to the atmosphere and back again—what variously has been described as "the pilgrimage of water" (NASA Science 2007: n.p.) or "the perpetual flux and exchange of water" (Marshall 2013: 1) (Figure 6.1). The origins of the hydrologic cycle are generally traced to the writing (and drawing) of Robert E. Horton (1933), whose research led to a broader classification of processes of infiltration, evaporation, interception and transpiration. The genius of Horton's hydrologic cycle lies in its simplicity: it is an abstraction that strips out all the complicating factors in order to understand certain key aspects of hydrology. However, in stripping out those factors, it also becomes an ideologically loaded representation of the environment, one that obscures as much as it reveals.

Perhaps the most important "complicating factor" stripped out of the model is human society. In the hydrologic cycle, water embarks on its various journeys and transformations without any people getting in the way—it simply evaporates, condenses and precipitates. To give just one example of why such a simplification fails, as we write this book, much of Europe is experiencing one of the worst droughts in recent decades. The hydrologic cycle might take us a little way in explaining why parched soils prevent infiltration, but it does not get us very far. To make sense of the drought, we need to consider a series of politicised processes: first, what is the role of human-induced climate change? Second, what is the impact of the investment

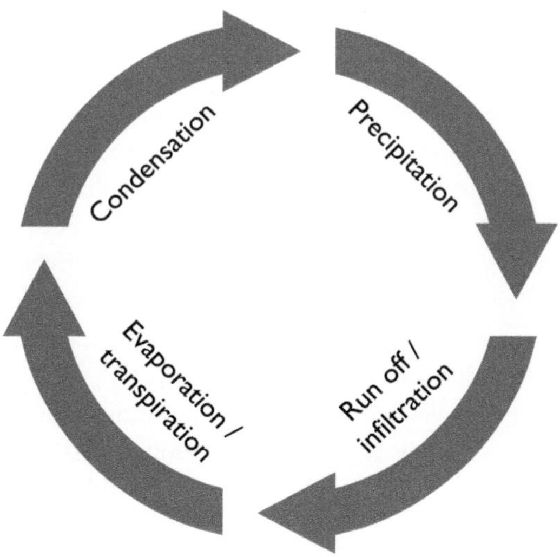

FIGURE 6.1 Simplified representation of the hydrologic cycle

decisions taken by states and business in recent years? And third, why is so much water used by some, while others struggle to access enough on which to live? When utilities in England and Wales[1] enforced a hosepipe ban in August 2022—restricting customers from watering their gardens to reduce pressure on the groundwater reserves—the industry regulator, Ofwat, reported that the water companies themselves lost an average of 2,923.8 million litres of water a day from leaky pipes in acute need of maintenance (Salvidge 2022). To explain the drought, therefore, political ecologists will emphasise the power relations and the inequalities that often are obscured or ignored by models, such as the hydrologic cycle.

The interdisciplinary concept of the "hydrosocial cycle" is one attempt to develop an alternative analytical tool that attends to these questions of power and scale. This concept politicises the flow of water and emphasises that, rather than a simple "pilgrimage", water is drawn to specific people and places by such things as money, the state and business. Jamie Linton and Jessica Budds (2014) are the two authors most closely associated with the term of the hydrosocial cycle. In calling for a shift in focus "from the hydrologic cycle to the hydrosocial cycle", they begin with a critique of the hydrologic cycle itself: they argue that the model should be viewed as part of a broader effort on the part of Horton and others to shape an emerging discipline of hydrology. Nevertheless, this emerging discipline, and the conceptualisations from which it emerged, are based on a largely Western worldview that fundamentally separates nature from society. It is no coincidence, therefore, that most visualisations of the hydrologic cycle neglect to include any evidence of social influences. Echoing Donna Haraway (1991), Linton and Budds argue that the hydrologic cycle is "partial and situated" in perspective (171).

The hydrosocial cycle represents a "relational-dialectical" approach to water. Water and society, Linton and Budds argue, co-constitute one another through their internal relations. Such a perspective contrasts with environmentally deterministic approaches that see the connection as one-way—that the hydrologic characteristics of an environment make people and societies who and what they are (Chapter 4). Often laden with racist pseudoscience, environmental determinism has at times served to justify imperial projects based on what is claimed to be a "civilising mission" (Arnold 2006). Interpretations of water and society have been similarly influenced by such problematic claims. In one of the most well-known works of environmental determinism, Ellen Semple (1911) develops an "anthropo-geography of rivers", supposedly demonstrating how different rivers, valleys and marshes all influence the societies of which they are a part. For Semple, lakes become "the nuclei of states" with a peculiar power inherent in them to unite peoples. Elsewhere, Karl Wittfogel (1957) developed what became an infamous thesis on "oriental despotism". Wittfogel's claim was that the availability of water resources across many Asian societies had given rise to a need for complicated irrigation projects that necessitated complex bureaucracies. According to this thesis, despotic regimes in Asia could be traced back to the particular ways in which water had to be managed.

Rejecting environmentally deterministic reasoning, some political ecologists—alongside environmental historians and other scholars—have emphasised the importance of human impacts *on* the environment. The Aral Sea was once the world's fourth largest lake, but since the 1960s, it has been shrinking so fast that by the late 1980s it had split in two. To intensify agricultural production in the Central Asian Soviet republics, Stalin ordered the massive construction of irrigation canals in the Uzbek SSR (today's Uzbekistan), moving water away from the Amu Darya and Syr Darya rivers that otherwise drained into the Aral Sea. Writing of the fervour to master Nature for economic development, the historian Paul Josephson (1995: 532–533) writes:

> Visionaries now proposed transforming lakes, ponds, rivers, and forests into "natural" electrical energy and agricultural factories. Rivers would be "regulated" for irrigation and electricity, "narrowed and straightened, destroying the shoals," ravines and marshes would be drained or filled, giving way to a system of orderly ponds and reservoirs. … Vast irrigation ditches stretched to the horizon with cheap electricity pumping water to formerly arid, unproductive land. … Lest anyone doubt that these were Stalin's projects, the tempo and scale of construction—armies of workers and equipment—far exceeded those in the West.

The environmental disaster of the Aral Sea did not give rise to an authoritarian regime; rather, the draining of the sea was made possible by the peculiar form of governance within the Soviet Union and the decisions made by an authoritarian regime seeking to make the steppes bloom. Instead of focusing on how nature is shaping society, researchers thus focus on how humans have utterly reconfigured the distribution of water resources throughout history. These interpretations are often referred to as "human impact models". However, for Linton and Budds, neither environmental determinism nor human impact models are adequate. Just as with the interpretation of "urban metabolism" presented in Chapter 5, the hydrosocial cycle instead captures mutual relationships—the co-determination of nature and society, of humans and water.

In emphasising these co-determining relationships, the hydrosocial cycle directs attention towards questions of ontology—that is, what water *is* and what political ecologies it is a part of. In doing so, the simple assumption that water consists of two molecules of hydrogen and one of oxygen—H_2O—and that it can be measured simply in terms of the available volume, gives way to a more nuanced interpretation of how water can exist in many different forms. Questions of quality, and the different qualities of different waters, become crucially important. Simultaneously, questions around water are also always questions of politics and policy. Many have tried to capture the hydrosocial cycle graphically—for the first time, the US Geological Survey released a diagram in 2022 that depicted not only "natural" aspects of the "water cycle" but also human water use—yet one image that stands out is Erik Swyngedouw's swirling visualisation of the "production of socio-nature" (Figure 6.2). In *Liquid Power*, Swyngedouw (2015) attempts to show the different relationships and processes that make water what it is, and the swirling image displays how human bodies and societies internalise their relations with the making of water in its many different forms.

Linton and Budds' work, and the clarity of it, has helped ensure that the hydrosocial cycle has been taken up as an analytical tool in a range of disciplines and in relation to a variety of empirical realities. Notably, it has been used to theorise indigenous waters (Wilson 2014), hydrosocial territories (Boelens et al. 2016), the human right to water (Sultana and Loftus 2019) and more. Inspiring both artists and geomorphologists, the concept provides an interdisciplinary lens through which we can make sense of the world in a more holistic manner, but it also raises a new set of questions around human-environment relations. To demonstrate the utility of the concept, we can put it to work in relation to the political economy of water.

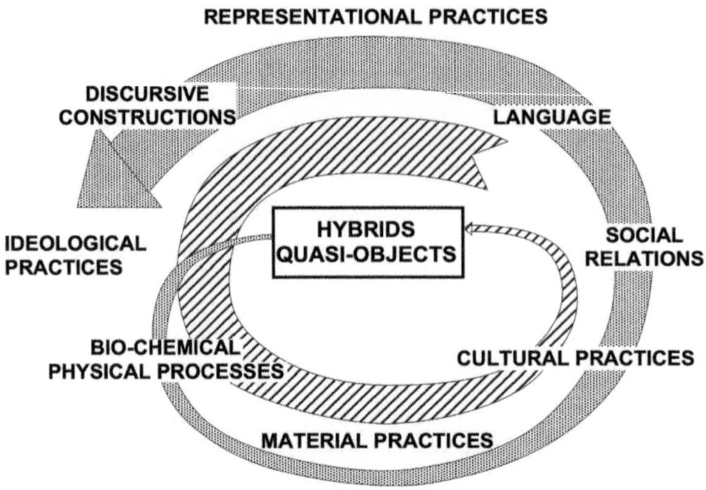

FIGURE 6.2 The hydrosocial cycle and the production of socio-nature

Figure by Erik Swyngedouw

POLITICAL ECONOMIES OF WATER

For many of us, the water we drink, cook with and shower in is likely to have been paid for by someone—maybe by you, your college, a landlord or a guardian. There is something a little awkward about the whole idea of paying for water that might grate with those who think of it as a common resource (Chapter 8), a gift from the gods or an intimate resource that makes us all "bodies of water". As we saw at the beginning of this chapter, many—and indigenous communities in particular—will refer to water as "life" or "lifeblood", but that does not prevent it from also being charged for in settler-colonial societies. If you were to speak to the CEO of a large water utility, the reason they would give for the need to pay for water is that it costs money to provide a service. Abstraction, purification, distribution and consumption all require infrastructure, energy, labour and investment. Nevertheless, the question of who provides that investment—and how they might recoup their initial outlays—has changed over time and across different geographical locations.

In the present, the "business of water", as we might refer to it, can be a case of tanker-truck drivers or water vendors seeking to find a niche market among poorer consumers in cities of the global south (Mustafa and Talozi 2018). Elsewhere, a business of water has emerged around small and medium-sized enterprises establishing water kiosks with whom major NGOs—such as Water Aid—work to ensure that piped water is distributed efficiently to poorer consumers (Rusca et al. 2015). For many consumers in the global north, the business of water instead entails stylish bottled water being carefully marketed to wealthier citizens according to mineral content, Total Dissolved Solids and effervescence (for a fascinating discussion of Total Dissolved Solids, see Acharya 2022).

The most lucrative business in potable water is perhaps found in the growth of large multinational utilities, often dominated by private equity firms, who seek to benefit from the provision of various services associated with piped water and sanitation. While global in reach and with a remarkable range of international investors involved, these utilities are the products of an extended history of water privatisation. This history notably originated in France but gained significant traction with the divestiture of water infrastructure in England and Wales in 1989. To make sense of a changing political economy of water, it is helpful to focus on the example of London—home to the most iconic of "financialised" water utilities, Thames Water. In this urbanised context, we need to be sensitive to the various power relations at work, as well as to how the local case can be understood in relation to processes and actors working on larger scales. The hydrosocial cycle is a particularly useful analytical tool to employ as it points us to a series of political ecological questions around who controls the flows of resources, who benefits from them and who loses out.

Private suppliers first developed a rudimentary water network in London in the seventeenth century, bringing water to wealthier residents of the city for a premium price. Excluded from this network, most residents instead relied on hand-dug wells or drew water directly from low-quality surface sources, such as the severely polluted tributaries running into the River Thames. The early private ventures in developing a water network proved utterly inadequate for the expanding urban population, and a municipal network was created in the late eighteenth century (Castro et al. 2003). Taxation initially provided public funding for this network; however, as the network grew—enabling water to be provided within the home—much of the cost of running, expanding and maintaining the network was passed on to individual households in the form of consumer charges. With a few exceptions, this remains the case today.

In England and Wales more broadly, consumer charges were long based on a fixed charge added together with a variable one. The latter was not connected with the metering of water use but the size and the location of the building (or, in more technical terms, the rateable value of a property). For much of the last two centuries, payment has therefore borne little relation to the amount of water consumed, as consumers instead have paid flat annual charges that make up the most significant part of the final bill. It makes some intuitive sense to charge on the basis of the rateable value of a property if we assume that (1) more people will live in a larger home; (2) they will possess a greater number of household utilities; (3) possibly also a garden; and (4) they as a result will use greater volumes of water. Interestingly, this charging system appears to do political work as it targets underlying socio-economic inequalities, assuming that richer people will occupy larger houses and therefore have to pay more.

In 1989, however, the system changed fundamentally. That year, the public utilities responsible for water provisioning in England and Wales were privatised, and the new private companies were able to set prices that reflected investments in the network and productivity increases in the utility. This price-setting was overseen by a consumer regulator (Ofwat) that could benchmark investments and price-increases in one utility against those in another. While metering of water consumption has been much less common in England and Wales than in other parts of the world, various plans for universal metering have recently also been pushed through across the network, although these have had varying levels of success (Loftus et al. 2016).

If both the ways of charging for water and the type of provider have varied historically within one location, such as London, they have also varied between locations. Thus, in France, rather than a publicly owned utility having been responsible for the construction and maintenance of a network, municipalities have signed concession contracts with private utilities for the running of services. Two of the largest global water utilities, Veolia and Suez, have therefore developed in this French context. These utilities not only operate in sectors ranging from entertainment to waste management in France, but as also the world saw a wave of privatisation unfolding in the early 1990s, they seized on the opportunity to bid for concessions to run water supplies around the world.

FROM PRIVATISATION TO RE-MUNICIPALISATION

Although there are key exceptions to the trend, notably in France and some cities in Spain (March 2015), municipal and state ownership of water utilities has been the dominant form of water provision throughout much of the last two centuries, not only in Europe but globally. This political economy of water has been interlaced with the "modern infrastructural ideal" we discussed in Chapter 5. However, the trend shifted in the 1990s following the privatisation of utilities in England and Wales under the government of Margaret Thatcher—an administration that sought to aggressively roll back state-involvement in sectors ranging from electricity and gas to housing and transportation. Water privatisation was nevertheless seen by many, both inside and outside Thatcher's government, as a more controversial move than other privatisations. As Veronica Strang (2004) argues, the intimate ways in which people relate to water—ways implied in the opening quote from Neimanis—made it seem a step too far. Beyond such personal objections, there were also other obstacles in the way that seemed to render water privatisation impossible.

First, in a free market, there is always a possibility of a company gaining monopoly control over the sector. In the case of water, such control is problematic and potentially dangerous: what, some might ask, if control over water were to end up in the "wrong" hands? Second, water itself is often considered to be a "natural monopoly". This means that unlike other utilities, such as electricity or gas, one provider tends to operate within a given location and distribute water through a geographically bounded network. Establishing market-based competition between providers is difficult when only one utility provides for a single network from which households receive their supplies. Third, without strong regulation, there is always a possibility that privatised utilities will over-exploit water resources as a way of generating higher profits.

Against these concerns—as well as fears over labour relations, foreign ownership, the stewardship of water resources and more—proponents in the global north argued that privatisation would be a solution to the large-scale investments needed in the water sector, something that the state appeared to be failing on. They also saw the market as an effective way of responding to the threat of water shortages, when lower supplies would mean higher prices, and market competition could also increase efficiencies within the formerly public sector. In the global south, arguments for privatisation also tended to focus on state failure. In Bolivia, for example, the state was deemed to have failed in its responsibility to provide for its citizens: the private sector was seen as the necessary alternative.

Jessica Budds and Gordon McGranahan (2003) summarise the differences in attitude towards water privatisation along a key conceptual fault line: is water essentially a public good or an economic good? Associated with this distinction is the question of whether democratic management of water resources, through the state, constitutes a better form of governance than using private actors acting in a regulated market economy. In 2000, the debates around privatisation culminated in highly significant protests in Bolivia's second largest city, Cochabamba. Known as the *Guerra del Agua* (Water War), the protests became deeply symbolic of a growing anger at what people variously referred to as corporate globalisation and neoliberalism—an anger reflected in further protests in Seattle in 1999 and Washington DC in 2000. Cochabamba's water was privatised through a concession contract led by the Californian energy giant Enron. As indigenous peasant farmers found themselves unable to access the groundwater supplies on which they depended, a highly organised urban working class demonstrated against having to pay increasing rates for water. Opposition to the government of Hugo Banzer and its neoliberal programme then came together in a series of protests, leading to Banzer's resignation and the eventual election of Bolivia's first indigenous president, Evo Morales, in 2005. As one slogan from the water war read: "Power to customary uses, long live water, die privatisers" (or in the original Quechua, "*Jallalla usos y costumbres, Kawsachun yaku, Wañachun privatizadores*") (quoted in Perreault 2006: 156).

Just as in the case of Cochabamba, water privatisation in the global south was heavily promoted by the International Financial Institutions—the World Bank and the IMF in particular—and it was often tied to loan conditionalities. Loans would be granted to nation states on condition that they began a process of economic reform, including the privatisation of water provision in key cities (Goldman 2005). With few examples of privatisation working in practice, the case of England and Wales together with experiences from Buenos Aires achieved an unusual prominence within the debates. Both cases also became the subject of early attempts by political ecologists to grapple with the emerging power relations—and the implications of scale—involved in water privatisation in a globalised political economy.

Focusing on England and Wales, Karen Bakker (2004) took the materiality of water as her starting point, arguing that water is an "uncooperative commodity". Given the ways it flows, water resists many of the attempts by economists and policymakers to turn it into a marketised product. For example, the expense and difficulty of transporting water from one catchment to another means that it is not easily exchangeable. There is, in other words, no way of coming around its material "watery" behaviour, which complicates the market dynamics that proponents of privatisation claim are necessary for the sector. Meanwhile, as new corporate actors became providers, privatisation seemed to unleash a process of deregulation, but Bakker shows that this simultaneously was a process of re-regulation. To guarantee that the providers acted in a politically acceptable way, the consumer regulator Ofwat was created, responsible for setting consumer prices, while the government-run Environment Agency would ensure the private utilities lived up to their environmental commitments. These regulators have had a necessary social and environmental influence through price capping and ensuring continuity of supply. However, as we return to below, Bakker also demonstrates how the newly privatised industry in a remarkable failure of governance was able to "produce scarcity", resulting in the Yorkshire drought of 1995.

Around the same time that Bakker's research pointed to the complexities of water provision in England and Wales, Alex Loftus and David McDonald (2001) developed a political ecological critique of water privatisation in Buenos Aires. Following the actor-oriented perspective on power (Chapter 2), notably suggested by Raymond Bryant and Sinéad Bailey (1997), they analysed the shifting relations between different actors within the politicised environment of the Argentinian capital city, including businesses, the state, unions and NGOs. In contrast to the more celebratory assessments of the World Bank and the IMF, Loftus and McDonald argued that any expansions to the water network—which was one of the main claims of success by the scheme's proponents—had only been made possible by significant increases in consumer charges. These increases involved reneging on contractual obligations and a contract renegotiation led by a minister who would later be jailed for corruption. When the Buenos Aires concession contract collapsed in 2005, water privatisation in the global south also seemed to slow considerably.

Several reasons have been given for the dramatic slowdown in the rate of water privatisation. Protests, such as those seen in the *Guerra del Agua*, certainly had an influence. Related to the politicised environments in which they acted, some multinational utilities also became concerned on the back of such protests that they would damage their reputations if seen to profit from poorer residents' livelihoods. At the same time, many of the most attractive concession contracts—those from which the greatest profits could be extracted—had already been cherry-picked, and other concessions looked far less attractive. Resonating with Bakker's argument that water is an uncooperative commodity, many of the newly privatised utilities found it harder than they initially hoped to make profits from the sale of water. To provide water requires large upfront financial investments in infrastructure that gradually crumbles into the ground (Gupta 2018). With the rapid turnover times expected of capital, it was difficult for governments pursuing privatisation to match new investors with utilities that would accrue profits over a fifty- to a hundred-year lifespan. By the 2010s, as Satoko Kishimoto et al. (2015) demonstrate, the rate of water privatisation had gone into reverse. Re-municipalisation—the return of control over water to the public sector—had instead become a far more prevalent trend.

FINANCIALISATION

In spite of the move away from outright privatisation, novel financial mechanisms have made it possible to profit from the provision of water in new ways. Financialisation is a term that, as Brett Christophers (2015) notes, has become something of a buzzword. As such, Christophers means that the term is used widely but without any real sense of what people are actually referring to. Indeed, financialisation may well refer to a shift in the locus of profitmaking from the "real economy", where water moves from one place to another, to the "financial economy" in which it is the trade in financial products (such as "securities" and "derivatives") that generates profits. It may also refer to a range of new actors, from sovereign wealth funds to insurance companies, that become involved in the ownership and management of different industries and services. For Costas Lapavitsas (2013), financialisation defines a process of "profiting without producing". His argument is that profitmaking has become so disconnected from the "real economy" that it is entirely reliant on speculative gains. In such an understanding, the production of potable water matters very little for the profits companies like Thames Water, Veolia and Suez now generate. We would take some issue with such a framing, arguing that there is a more complicated relationship between the accrual of profits from producing water and the guaranteed revenue streams that emerge on the back of this process of production. Thames Water provides a useful lens through which these processes might be understood.

While the utility's corporate structure has changed considerably from the initial model of "shareholder capitalism" that Thatcher claimed to be promoting, Thames Water came into being when water services were privatised in England and Wales in 1989. The UK government initially prevented majority-share ownership of the utility by itself retaining a "golden share", yet by 2001, Thames Water had been acquired by the German multinational energy company RWE. Then, in 2006, a consortium led by the Australian investment bank Macquarie Group went on to buy the utility. Known for its innovative financial products, which it had developed around Australian road and airport infrastructure, the Macquarie Group pursued a new financial strategy in Thames Water. With the utility guaranteed an income from the payment of household bills, this revenue stream could be transformed into financial commodities through a process of "securitisation" (Allen and Pryke 2013; Bayliss et al. 2023).

Securities are commodities that can be bought, at different levels of risk, by investors acting on the financial markets. Essentially, an investor pays the utility a sum of money upfront and in return receives the revenue generated from consumers continuing to pay their bills. The risk the investor takes is that the utility defaults and is unable to deliver the financial product it sells. As part of this process, the levels of debt within the Thames Water corporate structure have ballooned. While dividend payments to investors have been maintained, they have often exceeded the corporate profits generated in a given year. The dividend payments have, therefore, further increased the "debt-equity relation" within the utility: essentially, the Macquarie Group has increased the burden of debt within Thames Water without having to bear that burden themselves (Bayliss 2017). Furthermore, throughout much of the late 2000s and early 2010s, Thames Water paid virtually nothing in corporation tax, owing to a structure that involves large amounts of offshoring and the use of tax havens in the Caribbean.

For Thomas Purcell et al. (2020), the utility's monopoly position in relation to water supplies in the Greater London and Thames Valley area enables it to capture "monopoly rents". The Macquarie Group's securitisation of revenue streams has also made it possible to capture new forms of rent—sometimes termed "absolute rents" in reference to the barrier they erect in relation to competition—through the sale of these financial commodities. Rather than Thames Water "profiting without producing", as Lapavitsas argues, Purcell et al. point to a complex relationship in which value, rent and finance are internally related to one another, which raises questions about whose priorities govern the management of water in London and who has the right to the city. If we were to draw a cup of water from a tap in London, we could trace in it a sprawling network of socio-ecological relations, extending from the local to the global. The water would connect us via the plumbing and Thames Water to the sovereign wealth funds of Abu Dhabi and China as well as to the pension funds of British Telecom workers and UK university lecturers. In Chapter 5, we saw Matthew Gandy (2002) arguing that the history of cities at the same time must be read as a history of water. Following his argument, we could trace within the cup a history of contemporary urbanised capitalism.

THE PRODUCTION OF SCARCITY

During the European drought of 2022, numerous commentators in the UK media questioned the role of privatisation in the water sector. Rather like the argument Bakker made 20 years previously, they noted that privatisation had generated conditions in which the water utilities were likely to underinvest in infrastructure in order to generate larger profits. When long dry periods affected the country, therefore, the infrastructure was unable to cope (e.g. Salvidge 2022). This process reflects what political ecologists might refer to as the "production of scarcity"—a phenomenon observed not only in the water sector but also in the global oil industry (Bridge and Wood 2010). As we discussed in Chapter 4, political ecologists have long opposed Malthusian claims around the link between population pressure and resource scarcity, and in relation to water, claims of scarcity can never be explained only in terms of a "local absence". They must be studied critically in relation to the larger-scale political economic arrangements that regulate the flow of water and, thereby, enable some to benefit from it while leaving others with none.

The Yorkshire drought of 1995 is one of the most serious droughts to have affected any region in the United Kingdom in the post-privatisation era (Figure 6.3). Bakker (2000: 5), in her groundbreaking work, argues that the drought must be interpreted as "the production of water scarcity in nature" for three reasons: first, the utility responsible for providing water in the region, Yorkshire Water Services, failed to model climatic patterns effectively in order to anticipate the extremely hot and dry summer. Second, it also failed to adequately forecast demand. This was particularly problematic seeing that the price cap system put in place by Ofwat, which incentivised utilities to delay infrastructure investments within a given pricing round, exacerbated the impact of higher-than-expected demand. Finally, the new corporate structure of the utility, which post-privatisation involved outsourcing and job shedding, led to poor decision-making. While triggered by a heatwave, Bakker concludes that within the regulatory regime in which water was managed, the "1995–96 water shortage in Yorkshire was cost-effective, and the drought was a least-cost solution to constraints on the hydrosocial cycle" (20). First and foremost, the production of scarcity resulted from a failure of governance.

FIGURE 6.3 The old packhorse bridge across the Baitings Reservoir revealed during drought conditions in Yorkshire, England

Photo by iStock.com/Philip Silverman

As noted, the model of water privatisation that came about in England and Wales played an important role in the promotion of water privatisation through World Bank-led structural adjustment programmes. Even so, it was far from the first attempt to introduce market forces in the water sector. One important case is Chile where various neoliberal experiments were embarked upon under the dictatorship of Augusto Pinochet. In 1981, Pinochet signed the Chilean Water Code into law, which introduced a market in water rights. In a detailed analysis, Budds (2004) takes two valleys in the semi-arid Norte Chico as her starting point to examine the implications of the Water Code. At the time, a massive expansion of avocado farming in the valley of La Ligua was creating growing pressures on groundwater sources in the region. Water scarcity became an issue, yet Budds demonstrates how this scarcity was actively produced, not only through intensive avocado farming but also through the role of speculators hoping for a rising price in water rights. Historically, land ownership in La Ligua has been organised in different ways. Under the socialist government of Salvador Allende (which was toppled by Pinochet), land was sold to peasant collectives. But this policy radically shifted under Pinochet who prioritised private land ownership. As a result, water scarcity was unevenly experienced when peasant smallholders experienced the lowest levels of water security. For Budds, a political ecology approach to the Water Code was essential because of the way it allowed her to move "towards more fundamental questions about who produces nature and with what social and ecological outcomes" (325). The production of scarcity was one of these outcomes.

WATER (IN)SECURITY: DROUGHTS AND FLOODS

One of the most common environmentally determinist arguments is the claim that the next world war will be over water. "Water wars" have figured prominently in the International Relations literature, often taking a cue from Robert Kaplan's (1994: 61) framing of the environment as "the national security issue of the early twenty-first century" (see notably Starr 1989 and Klare 2020 for influential accounts). More recently, arguments around water scarcity and its likely influence on human societies have centred on a notion of water security. Water security is framed differently by different groups: following Kaplan, it can be interpreted as a concern for national security in the face of heightening environmental threats. Within such a framing, the object to be secured is clearly the water resources within a given nation state, and the concern is for the security of that state. However, water security can also be framed as a question of human security. In this sense, the referent is not an abstract nation state but groups of people who may lack access to safe, sufficient water supplies in their neighbourhood or who may be exposed to the threat of flooding.

This latter framing challenges some of the cruder ways in which water has been described as a national security issue, but that determinism is still not completely absent. For example, for Casey Brown and Upmanu Lall (2006), water security refers to how the variability of rainfall poses a threat to economic prosperity. Political ecologists challenge these readings by placing the political economic relations that produce scarcity centre stage: a critical study should not only focus on the levels of precipitation but also identify the underlying power relations that give rise to a particular socio-ecological configuration, the outcome of which is water (in)security. Indeed, rather than vulnerable populations being referred to as "hostages to hydrology" (Grey

FIGURE 6.4 Floods in Halych, Ukraine in the summer of 2020

Photo by iStock.com/Bilanol

and Sadoff 2007: 557), a political ecology of water points to the production of scarcity, just as it developed in Yorkshire and La Ligua.

For most scholars concerned with water security, the risk to be addressed is not just that of too little water threatening livelihoods but also that of too much water, for example from flooding (Figure 6.4). Depending on the lens through which this hazardous phenomenon is analysed—fluvial geomorphology, climate science or insurance risk modelling, to mention a few such "lenses"—authors will emphasise different aspects of floods. As with the critiques of Malthusianism and natural scarcity, political ecologists tend to focus on the socio-ecological relations through which flooding and flood risk are produced. A key conceptual distinction here is that the risk of a disaster happening is proportional to two factors: first, the magnitude of the hazard, such as the flood or the drought; and second, the social vulnerability of the individual or the community being exposed to that hazard. As Ben Wisner et al. (2004: 11) argue, it is "the [classed, racialised, gendered, etc.] characteristics of a person or a group and their situation that influence their capacity to anticipate, cope with, resist and recover from the impact of a natural hazard". Floodwaters themselves cause no disaster, but the relation between that water and a vulnerable community is potentially calamitous. Thus, to borrow an expression from Neil Smith (2006: n.p.), political ecologists will often claim that "there's no such thing as a natural disaster".

Smith wrote these words only a few months after Hurricane Katrina had claimed the lives of well over a thousand residents of New Orleans in September 2005, mainly because of flooding. Smith was not in any way denying this awful loss of life—one that many consider to be one of the worst "natural disasters" in US history. Instead, his argument was that the events of September did not need to *become* a disaster: political and economic decisions made before and at the time actively created it. Echoing Wisner et al., Smith argues that if we are to get to the root causes of catastrophic flooding, we must "denaturalise" natural disasters. He also points to several of these root causes in the case of Katrina. The racialised geographies of urban space and topography in New Orleans ensured that wealthier white residents lived in districts far less likely to flood than people of colour. Decisions to construct levees on the Mississippi River—and the hubris of the army corps of engineers—created the possibility that the levees would be breached, thereby flooding lower-lying areas of the city. Smith also notes that the failures of the Federal Emergency Management Agency, run by a political appointee of the Bush administration, resulted in a lack of disaster preparedness. A poor evacuation plan, and a neglect of those predominantly people of colour who were unable to evacuate, further exacerbated the loss of life. In short, those who died in the wake of Hurricane Katrina need not have died. Just as the disaster resulted from torrential rains, it cannot be explained without also taking socio-political processes and decision-making into consideration.

Smith's intervention is written with a deep sense of anger at the needless loss of life in New Orleans. It is also clear that he saw little hope in the federal administration of the time learning from the disaster. Instead, the likely "naturalisation" of a market-based response would further heighten vulnerabilities to flooding. However, he does find a glimmer of hope in the grassroots groups in New Orleans who might "build back better". In a very different context—the Andaman and Nicobar Islands in the Indian Ocean—Sophie Blackburn (2018) points to the ways in which the response to a flood event could be transformational. On 26 December 2004, a powerful tsunami struck the Andamans, following an earthquake in the Indian Ocean. Waves of up to 30 metres engulfed coastal areas throughout the region with parts of Indonesia, Thailand and Sri Lanka being particularly badly affected. In some respects, Blackburn picks up where Smith leaves off. She begins by asking what possibilities there are for doing things differently after a disaster on the scale of the Indian Ocean tsunami. The large teams of NGO and relief workers that arrived in the Andamans could end up reproducing a problematic model of Western development as "deliberate transformation". Instead, she sees the possibility for a community-driven kind of transformation, building on rights, to reduce vulnerability and generate new political outcomes. Blackburn's is, therefore, one of several approaches that can be taken up by political ecologists in striving for water security—one not based on business-as-usual but a transformative political project centred on human rights.

WATER JUSTICE AND THE HUMAN RIGHT TO WATER

In 2010, the UN General Assembly formally recognised the human right to water. Some states, such as South Africa, had already recognised a human right to water in their constitutions, but with UN Resolution 64/292, new legal possibilities for citizens to place demands on their governments were created. For some, the resolution signalled a positive shift in water governance, one which would make it possible to achieve fairer and more democratic access arrangements around water (Barlow 2012). Others, however, disagreed. Bakker (2013) suggested that the "right to water" risked being understood as a right to private ownership, on a par with the "water rights" granted in Pinochet's Chile, thereby making it possible to exclude others from usage. Rather than water being a common pool resource (Chapter 8), the discussion of water rights blurred with property rights, implying formalised individual ownership instead of a shared democratic project. Upon reflection, there is nothing within the recognition of a human right to water that prevents further private sector involvement in the provision of supplies. This view seemed to be confirmed when one of the main mouthpieces for the private sector, *Global Water Intelligence*, only a day after the UN resolution wrote that the United Nations' decision should be viewed "as a massive *defeat* for the global water justice movement" (Gasson 2010: n.p., emphasis added).

South Africa's constitutional recognition of the human right to water helps to illustrate why such a pledge might be read as both a victory and a defeat. Adopted during the transition from apartheid, the Constitution of 1996 is a remarkably progressive document that captures the ambitions of what many referred to as "the rainbow nation" (though see Habib 1997 for a critical account). The constitution's Bill of Rights guarantees that "everyone has the right to have access to ... sufficient food and water", and it issues that "the state must take reasonable legislative and other measures within its available resources to achieve the progressive realisation of these rights" (chap. 2, sect. 27:1). Nevertheless, when the Human Sciences Research Council of South Africa conducted a hotly debated survey in the early 2000s, it found that up to 10 million South Africans were being affected by water disconnections for the non-payment of bills (McDonald 2002). The human right to water, then, seemed to have little bearing on whether a household would have continued access to water if they were unable to pay.

The right of South Africa's municipalities to disconnect water supplies for non-payment has been challenged in several court cases, but it has repeatedly been reasserted. For the courts, the utilities' right to recover all costs from the provision of a water service trumps the right to water pledged within the constitution. The two rights are in conflict with each other. Moreover, the installation of pre-payment meters in some poorer communities, requiring residents to charge up a pre-payment card in order to access water, only seems to have reinforced the concern that the right to water has little material bearing on water access. And yet, in the complex textures of everyday life, it is fascinating to see how residents have come up with creative technical solutions to bypass the pre-payment meters in resistance to the political economy they enact (von Schnitzler 2013).

Focusing on the human right to water in Bolivia, Rocio Bustamante et al. (2012) argue that attempts to assert this right easily end up privileging one institutional form—the state—over and above others. Indeed, classical liberal state theory holds that the state first developed to guarantee the rights of the individual in relation to others. However, as we saw in Chapter 3, the development of the modern state can also be traced through a long history of European territorial governance and colonisation, and in a settler-colonial context, a focus on "rights", therefore, means privileging the very institutions that indigenous groups often see as illegitimate (Temper 2016). These arguments are crucially important to consider if we seek to displace an Anglo-American gaze that so often sets the norm in political ecology and other critical environmental studies. For us, it is still important to recognise the "human right to water" as a rallying cry around which different groups mobilise for water justice. Rather than an endpoint, the right to water must be viewed as a process; a concept around which movements for water justice coalesce and give greater meaning to their practices. To read the right to water in this manner allows movements to work with, against but also beyond the institutional form of the state, seeking new ways of producing the hydrosocial cycle (Angel and Loftus 2019).

CONCLUSION AND LEARNING OUTCOMES

Water is vital to all life. As Astrida Neimanis argues, we are as humans in part water. In this chapter, we have identified the hydrosocial cycle as a key analytical concept for political ecologists. This concept enables us to challenge but also transcend the separation of humans from their broader environments, a separation implied by scientific models, such as the hydrologic cycle. The hydrosocial cycle draws attention to the ways in which human practices influence the flow of water, and how those flows, in turn, come to influence the societies of which they are a part. For example, lived experiences of water scarcity must from this perspective be seen as the outcomes of particular political economic configurations, even as the "wateriness" of water has a profound influence on those configurations.

Across several historical-geographical contexts, we have witnessed how different forms of ownership, corporate structures and grassroots movements mediate the flow of water, including struggles around state control, privatisation and financialisation. By exploring the influence of these processes on the hydrosocial cycle, a rich political ecology literature demonstrates how power relations make it possible for some to generate super-profits while producing water scarcity and hazardous environments for others. It is tempting to see water insecurity as a phenomenon determined by naturally occurring processes, but just as water scarcity, it needs to be understood as a socially produced phenomenon. Water security, water justice and the human right to water can, therefore, be understood as vehicles for securing more democratic access to water.

Having read this chapter, we hope that you can:

- Apply the hydrosocial cycle as a political ecological concept
- Account for some of the changing political economies through which networked water supplies have been provided
- Explain financialisation as a global process that reworks local political economies of water
- Distinguish between different uses of the term water security, especially in the contexts of water scarcity (drought) and overabundance (flooding)
- Evaluate the debates surrounding the human right to water

FURTHER READING

Bakker, Karen. 2013. "Neoliberal versus Postneoliberal Water: Geographies of Privatization and Resistance." *Annals of the Association of American Geographers* 103(2): 253–260.

Bakker is one of the most prolific political ecologists working on water, and this article provides excellent further reading on the history of water privatisation as well as resistance to it around the world.

Linton, Jamie, and Jessica Budds. 2014. "The Hydrosocial Cycle: Defining and Mobilizing a Relational-Dialectical Approach to Water." *Geoforum* 57: 170–180.

This is a key text on the hydrosocial cycle, defining it as a framing concept for a political ecology of water. Written in clear prose with informative illustrations, it is a great article to read after this chapter to gain a more finely textured understanding of the processes by which water and society inform each other.

NOTE

1. When referring to the United Kingdom in this chapter, we will make a distinction between water governance in England and Wales, where water utilities were privatised in 1989, and water governance in Scotland, where water privatisation was resisted. The utility Scottish Water remains fully publicly owned.

BIBLIOGRAPHY

Acharya, Amitangshu. 2022. "Political Ecology of Small Things." PhD thesis, University of Edinburgh, United Kingdom.

Allen, John, and Michael Pryke. 2013. "Financialising Household Water: Thames Water, MEIF, and 'Ring-Fenced' Politics." *Cambridge Journal of Regions, Economy and Society* 6(3): 419–439.

Angel, James, and Alex Loftus. 2019. "With-against-and-beyond the Human Right to Water." *Geoforum* 98: 206–213.

Arnold, David. 2006. *The Tropics and the Traveling Gaze: India, Landscape, and Science, 1800-1856.* Seattle, WA: University of Washington Press.

Bakker, Karen J. 2000. "Privatizing Water, Producing Scarcity: The Yorkshire Drought of 1995." *Economic Geography* 76(1): 4–27.

Bakker, Karen J. 2004. *An Uncooperative Commodity: Privatizing Water in England and Wales.* Oxford, UK: Oxford University Press.

Bakker, Karen. 2013. "Neoliberal versus Postneoliberal Water: Geographies of Privatization and Resistance." *Annals of the Association of American Geographers* 103(2): 253–260.

Barlow, Maud. 2012. "Foreword." In *The Right to Water: Politics, Governance and Social Struggles*, edited by Farhana Sultana and Alex Loftus, xv–xvii. London, UK: Earthscan.

Bayliss, Kate. 2017. "Material Cultures of Water Financialisation in England and Wales." *New Political Economy* 22(4): 383–397.

Bayliss, Kate, Elisa Van Waeyenberge, and Benjamin O. L. Bowles. 2023. "Private Equity and the Regulation of Financialised Infrastructure: The Case of Macquarie in Britain's Water and Energy Networks." *New Political Economy* 28(2): 155–172.

Blackburn, Sophie. 2018. "What Does Transformation Look Like? Post-Disaster Politics and the Case for Progressive Rehabilitation." *Sustainability* 10(7): 2317.

Boelens, Rutgerd, Jaime Hoogesteger, Erik Swyngedouw, Jeroen Vos, and Philippus Wester. 2016. "Hydrosocial Territories: A Political Ecology Perspective." *Water International* 41(1): 1–14.

Bridge, Gavin, and Andrew Wood. 2010. "Less Is More: Spectres of Scarcity and the Politics of Resource Access in the Upstream Oil Sector." *Geoforum* 41(4): 565–576.

Brown, Casey, and Upmanu Lall. 2006. "Water and Economic Development: The Role of Variability and a Framework for Resilience." *Natural Resources Forum* 30(4): 306–317.

Bryant, Raymond L., and Sinéad Bailey. 1997. *Third World Political Ecology*. London, UK: Routledge.

Budds, Jessica. 2004. "Power, Nature and Neoliberalism: The Political Ecology of Water in Chile." *Singapore Journal of Tropical Geography* 25(3): 322–342.

Budds, Jessica, and Gordon McGranahan. 2003. "Are the Debates on Water Privatization Missing the Point? Experiences from Africa, Asia and Latin America." *Environment and Urbanization* 15(2): 87–114.

Bustamante, Rocio, Carlos Crespo, and Anna M. Walnycki. 2012. "Seeing through the Concept of Water as a Human Right in Bolivia." In *The Right to Water: Politics, Governance and Social Struggles*, edited by Farhana Sultana and Alex Loftus, 223–240. London, UK: Earthscan.

Castro, José E., Maria Kaika, and Erik Swyngedouw. 2003. "London: Structural Continuities and Institutional Change in Water Management." *European Planning Studies* 11(3): 283–298.

Christophers, Brett. 2015. "The Limits to Financialization." *Dialogues in Human Geography* 5(2): 183–200.

Gandy, Matthew. 2002. *Concrete and Clay: Reworking Nature in New York City*. Cambridge, MA: MIT Press.

Gasson, Christopher. 2010. "The Human Right to a National Water Plan." *Global Water Intelligence*, 29 July. https://www.globalwaterintel.com/news/2010/30/the-human-right-to-a-national-water-plan. Accessed: 27 September 2022.

Goldman, Michael. 2005. *Imperial Nature: The World Bank and Struggles for Social Justice in the Age of Globalization*. New Haven, CT: Yale University Press.

Grey, David, and Claudia W. Sadoff. 2007. "Sink or Swim? Water Security for Growth and Development." *Water Policy* 9(6): 545–571.

Gupta, Akhil. 2018. "The Future in Ruins: Thoughts on the Temporality of Infrastructure." In *The Promise of Infrastructure*, edited by Nikhil Anand, Akhil Gupta, and Hannah Appel, 62–79. Durham, NC: Duke University Press.

Habib, Adam. 1997. "South Africa—the Rainbow Nation and Prospects for Consolidating Democracy." *African Journal of Political Science/Revue Africaine de Science Politique* 2(2): 15–37.

Haraway, Donna. 1991. *Simians, Cyborgs, and Women: The Reinvention of Nature*. London, UK: Free Association Books.

Horton, Robert E. 1933. "The Role of Infiltration in the Hydrologic Cycle." *Eos, Transactions, American Geophysical Union* 14(1): 446–460.

Josephson, Paul R. 1995. "'Projects of the Century' in Soviet History: Large-Scale Technologies from Lenin to Gorbachev." *Technology and Culture* 36(3): 519–559.

Kaplan, Robert. 1994. "The Coming Anarchy: How Scarcity, Crime, Overpopulation, Tribalism, and Disease Are Rapidly Destroying the Social Fabric of Our Planet." *The Atlantic*, February. https://www.theatlantic.com/magazine/archive/1994/02/the-coming-anarchy/304670. Accessed: 27 September 2022.

Kishimoto, Satoko, Emanuele Lobina, and Olivier Petitjean. 2015. *Our Public Water Future: The Global Experience with Remunicipalisation*. Amsterdam, Netherlands: Transnational Institute (TNI)/Public Services International Research Unit (PSIRU)/Multinationals Observatory/Municipal Services Project (MSP)/European Federation of Public Service Unions (EPSU).

Klare, Michael T. 2020. "Climate Change, Water Scarcity, and the Potential for Interstate Conflict in South Asia." *Journal of Strategic Security* 13(4): 109–122.

Lapavitsas, Costas. 2013. *Profiting without Producing: How Finance Exploits Us All*. London, UK: Verso.

Lewis, Sophie. 2018. *Full Surrogacy Now: Feminism against Family*. London, UK: Verso.

Linton, Jamie, and Jessica Budds. 2014. "The Hydrosocial Cycle: Defining and Mobilizing a Relational-Dialectical Approach to Water." *Geoforum* 57: 170–180.

Loftus, Alex, Hug March, and Fiona Nash. 2016. "Water Infrastructure and the Making of Financial Subjects in the South East of England." *Water Alternatives* 9(2): 319–335.

Loftus, Alexander J., and David A. McDonald. 2001. "Of Liquid Dreams: A Political Ecology of Water Privatization in Buenos Aires." *Environment and Urbanization* 13(2): 179–199.

March, Hug. 2015. "Taming, Controlling and Metabolizing Flows: Water and the Urbanization Process of Barcelona and Madrid (1850–2012)." *European Urban and Regional Studies* 22(4): 350–367.

Marshall, Shawn J. 2013. "Hydrology." *Elsevier Reference Module in Earth Systems and Environmental Sciences*. https://www.doi.org/10.1016/B978-0-12-409548-9.05356-2.

McDonald, David A. 2002. "The Bell Tolls for Thee: Cost Recovery, Cutoffs, and the Affordability of Municipal Services in South Africa." In *Cost Recovery and the Crisis of Service Delivery in South Africa*, edited by David A. McDonald and Pape John, 161–179. Cape Town, South Africa: Human Sciences Research Council Publishers.

Mustafa, Daanish, and Samer Talozi. 2018. "Tankers, Wells, Pipes and Pumps: Agents and Mediators of Water Geographies in Amman, Jordan." *Water Alternatives* 11(3): 916–932.

NASA Science. 2007. "Water Cycle." https://science.nasa.gov/earth-science/oceanography/ocean-earth-system/ocean-water-cycle. Accessed: 27 September 2022.

Neimanis, Astrida. 2013. "Feminist Subjectivity, Watered." *Feminist Review* 103(1): 23–41.

Perreault, Thomas. 2006. "From the *Guerra del Agua* to the *Guerra del Gas*: Resource Governance, Neoliberalism and Popular Protest in Bolivia." *Antipode: A Radical Journal of Geography* 38(1): 150–172.

Purcell, Thomas F., Alex Loftus, and March Hug. 2020. "Value-Rent-Finance." *Progress in Human Geography* 44(3): 437–456.

Rusca, Maria, Klaas Schwartz, Lejla Hadzovic, and Rhodante Ahlers. 2015. "Adapting Generic Models through Bricolage: Elite Capture of Water Users Associations in Peri-Urban Lilongwe." *The European Journal of Development Research* 27(5): 777–792.

Salvidge, Rachel. 2022. "Water Firms in England and Wales Lost 1tn Litres via Leaky Pipes in 2021." *The Guardian*, 19 August. https://www.theguardian.com/environment/2022/aug/19/water-firms-england-wales-litres-leaky-pipes-ofwat. Accessed: 24 August 2022.

Sands, Leo. 2022. "Pakistan Floods: One Third of Country Is Under Water—Minister." *BBC News*, 30 August. https://www.bbc.com/news/world-europe-62712301. Accessed: 27 September 2022.

Semple, Ellen. 1911. *Influences of Geographic Environment: On the Basis of Ratzel's System of Anthropo-Geography*. New York, NY: Henry Holt & Co.

Smith, Neil. 2006. "There's No Such Thing as a Natural Disaster." *Understanding Katrina: Perspectives from the Social Sciences* 11. https://items.ssrc.org/understanding-katrina/theres-no-such-thing-as-a-natural-disaster. Accessed: 27 September 2022.

Starr, Joyce R. 1989. "Water Wars." *Foreign Policy* 82: 17–36.

Strang, Veronica. 2004. *The Meaning of Water*. London, UK: Routledge.

Sultana, Farhana, and Alex Loftus, eds. 2019. *Water Politics: Governance, Justice and the Right to Water*. London, UK: Earthscan.

Swyngedouw, Erik. 2015. *Liquid Power: Contested Hydro-Modernities in Twentieth-Century Spain*. Cambridge, MA: MIT Press.

Temper, Leah. 2016. "Blocking Pipelines, Unsettling Environmental Justice: From Rights to Nature to Responsibility to Territory." *Local Environment* 24(2): 94–112.

von Schnitzler, Antina. 2013. "Traveling Technologies: Infrastructure, Ethical Regimes, and the Materiality of Politics in South Africa." *Cultural Anthropology* 28(4): 670–693.

Wilson, Nicole J. 2014. "Indigenous Water Governance: Insights from the Hydrosocial Relations of the Koyukon Athabascan Village of Ruby, Alaska." *Geoforum* 57: 1–11.

Wisner, Ben, Piers Blaikie, Terry Cannon, and Ian Davies. 2004. *At Risk: Natural Hazards, People's Vulnerability and Disasters*, 2nd ed. London, UK: Routledge.

Wittfogel, Karl A. 1957. *Oriental Despotism: A Comparative Study of Total Power*. New Haven, CT: Yale University Press.

CHAPTER 7

Energy

The hottest summer in Swedish history was coming to an end when Greta Thunberg began her school strike outside the national parliament. Large wildfires had spread in the country's forested landscape; people had been evacuated, animals died and the timber industry faced massive capital destruction in the flames. The summer of 2018 also marked another occasion: global carbon dioxide (CO_2) emissions from fossil fuel combustion reached an all-time high. Half of the emissions resulted from electricity and heat generation; one-quarter came from transportation (IEA 2020). While we cannot attribute a single wildfire or other natural hazard to climate change, more intensive disasters are occurring each year at a time when the global mean temperature is rising (Coppola 2021). In Stockholm, Thunberg called for climate action. Four years later, the global school strike movement reported that 17 million students in 8,600 cities had joined her in protest (Fridays for Future 2022).

Enveloped in the strikers' demands is a concern with energy. Energy production, fossil fuel dependence and the prospects of a low-carbon energy transition have also become some of the most topical areas of research in political ecology over the past decade. Two main ways of approaching energy have emerged: first, as something *embedded* in social, economic and political processes, and second, as something *extracted* from nature for human benefit. In this chapter, we explore both approaches, but first, we examine an underlying, more fundamental question with surprisingly far-reaching implications. The question is simple: what is energy?

WHAT IS ENERGY?

The social critic Ivan Illich (2010) once observed that energy is a modern idea with a short history. "Energy", he argued, "has become something that individuals and societies need. It is a symbol that fits our age, the symbol of that which is both abundant and scarce" (13). As Figure 7.1 shows, global energy supplies have increased to unprecedented levels over the past three decades, but even so, wars are fought over energy resources as if they were scarce, the phenomenon of "peak oil" raises fears that the global economy is running out of fuel, and renewable energy appears difficult to harness in sufficient quantities. The threat of an energy crisis, as Gavin Bridge (2011: 308) notes, "is the existential terror of an empty tank and flaccid fuel lines: modernity, fused". Paradoxically, there seems to be both too much and too little energy around.

DOI: 10.4324/9781003095521-7

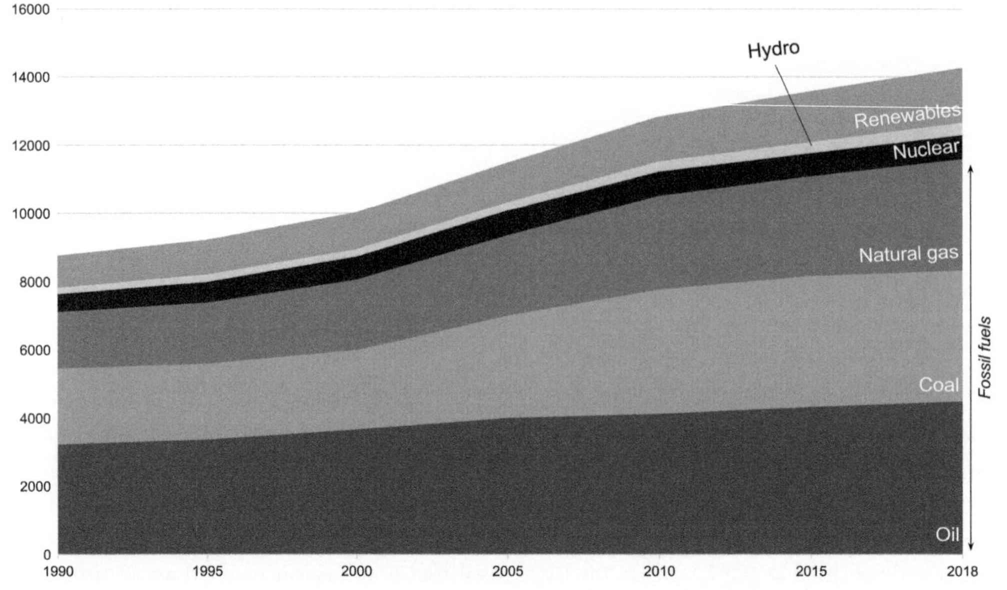

FIGURE 7.1 World total energy supply by source, 1990–2018 (unit: Mtoe)

The graph shows that hydropower has had the highest growth rate (97 per cent) over the past three decades, although hydropower still only contributed 2.5 per cent of the total energy supply in 2018. While an important aspect of a low-carbon energy transition is to reduce the combustion of fossil fuels, the supplies of oil, coal and natural gas have steadily increased on a global scale (with 39, 73 and 96 per cent, respectively). In 2018, fossil fuels made up 81 per cent of the world's total energy supplies.

Based on IEA data from IEA (2020) World Energy Balances 2020, www.iea.com/statistics, All rights reserved; as modified by the Authors.

To unravel this paradox, we can trace the origins of the energy idea. In the mid-1800s, Glasgow had become a hub of imperial shipping, and it was known widely as the "Second City of the British Empire". Raw cotton arrived in the harbour from slave plantations in the American South, and during the American Civil War increasingly from India and Egypt, to be milled and woven into fabric. The novelty of the time was the thrumming of the steam engines and the power looms in the city's factories, a novelty that some attribute to the starting point of the "Great Acceleration" and the Anthropocene (McNeill and Engelke 2016). To operate a steam engine, you shovel coal in on one end; you put it on fire, and on the other, you get a piston moving which brings the loom to life. As a by-product, you also get different kinds of waste—heat energy and CO_2 not least. For the engineers working with the steam engine, the machine did something puzzling: it converted a lump of coal into motion. If you think about it, coal and motion are two very different things. At the same time, this appeared to be a one-way process. Coal could generate motion, but the moving piston could not generate coal. For the engineers, it was important to explain these conversions in order to maximise the work done by the machine using as little coal as possible. The theory they developed became the science of thermodynamics, which is a cornerstone of modern physics, chemistry and ecology (Kondepudi and Prigogine 1998; Jørgensen 2012).

The engineers developed a set of thermodynamic "laws" to describe what energy is and what it does. The First law of thermodynamics states that energy is something you can convert from one form to another but never deplete. For instance, thermal energy can become mechanical energy in a steam engine, electricity can metabolise a city (Chapter 5) and the chemical energy stored in food can become movement, heat and thoughts in a human body. The peculiar thing is that none of these energy forms is "energy" in and by itself; it is a particular manifestation of it, and the total amount of "energy" always stays the same. The physicist Richard Feynman (2013: n.p.) famously explained that energy "is a most abstract idea, because it is a mathematical principle":

> it says that there is a numerical quantity which does not change when something happens. It is not a description of a mechanism, or anything concrete; it is just a strange fact that we can calculate some number and when we finish watching nature go through her tricks and calculate the number again, it is the same.

The scientific concept of energy is an elusive one—a mathematical abstraction.

The Second law of thermodynamics is even more intriguing than the first. It explains why energy conversions are always one-way. The so-called Entropy law states that as soon as energy changes form, it is degraded. It loses in quality, or more specifically, it becomes increasingly disorderly. The degree of disorder is easy to picture in the case of a lump of coal. When you put it on fire, you release the energy stored in it, and while this gets the steam engine running, you also get all of the by-products. Disorder, or "entropy", increases in the energy system. This means that to keep a steam engine running, you need to provide it with a continuous supply of low entropy. If you fail to add more, the process grinds to a halt. And this also goes for a lightbulb, a car, an internet server, a human body, a wildfire, an ecosystem, a city and a global economy. The Entropy law, therefore, gives us one part of the reason why fossil fuels have become such an integral part of the industrial economy: few things can measure up to a fossil fuel in terms of supplying low entropy while being easy to transport to a location where it is in demand.

ENERGY AND THE COLONIALITY OF KNOWLEDGE

Political ecologists have recently taken an interest in the history of thermodynamics as a way of describing the environment. This interest ties into a wider effort to understand how ecological science and politics interact on an epistemological level. Focusing on the close relationship between knowledge and power (Chapter 2), Tim Forsyth (2015) argues that we need to go beyond common distinctions that place science and politics into separate domains. Instead, we need to appreciate how the production of knowledge always takes place in a historical context that embeds social priorities, cultural norms and value-based assumptions. The context within which a knowing subject is positioned influences the aims of inquiry, the questions it seeks to answer as well as the selection and interpretation of data. As noted in previous chapters, Donna Haraway (1991) argues that knowledge always is "situated" in social relations and experiences of gender, class and race. Inescapably, knowledge is partial in perspective. Claims to universal knowledge are, therefore, ultimately claims that seek to establish a particular outlook on the world as "natural" and "objective", and this is usually the outlook of the powerful. Thus, when we make claims to knowledge, we must be reflexive about where we know *from*. As opposed to a disembodied gaze from nowhere, Haraway writes, "partial perspective can be held accountable" (190).

In *The Birth of Energy*, Cara New Daggett (2019) argues that the science of thermodynamics cannot be separated from the social and economic context from which it emerged. The questions the engineers asked about how to maximise the work efficiency of the steam engine inherently reflected the driving force behind the capitalist economy in which the machine was used: the profit motive. Coal was always in short supply and came at a high price, and with increasing efficiency, factory owners could generate larger profits. Daggett also takes a further step to show how the concepts physicists and ecologists use to describe the natural world not only derive from human interaction with "nature" but also the cultural frames of reference they have to interpret it. Beyond the profit motive, thermodynamics reflected key tenets of the engineers' Scottish Presbyterian faith. According to the First law of thermodynamics, "energy" was a constant—like God—but following the Second, its usefulness declined as soon as the flawed humans commanding it put it to use. The laws of thermodynamics, therefore, encapsulated an ethical imperative that resonated with the engineers' Protestant cosmology: in a world of scarcity, the energy idea exalted hard work and condemned sin and sloth.

As an abstraction, the energy idea folds all instances of energy use into the same universal category. A rocket launch and a horse-cart ride—two entirely different cultural practices from the perspectives of the astronaut and the farmer—are subsumed under the same economistic logic of energy conversions and efficiencies. As Larry Lohmann (2016: 47) argues, there was no such thing as "energy" before the Industrial Revolution: "People talked about horses, fire, trade winds, lightning, the ripening of wheat and so forth", yet by the late nineteenth century, "'energy' had become part of the currency of 'nature'". As an environmental epistemology, therefore, the energy idea obscures the political, social and cultural significances of human-environment interactions, and as such, it is a powerful ideological construct. Energy is a way of representing the environment that derives from the work of European engineers in the core of a global empire.

The energy idea provides a method for gaining knowledge about nature and points to certain courses of action. At the same time, it contests other ways of knowing the world. "The claim that energy is universal", Daggett (2019: 69) argues, "can also be understood as a Western claim that a certain epistemology … is universal". Figure 7.2 shows a map of South America which we have taken from the Global Solar Atlas. The World Bank has developed this mapping tool with the stated aim of supporting "the scale-up of solar power in our client countries" (GSA 2021). One of the world's largest creditors, the World Bank is also one of the International Financial Institutions that have contributed the most to environmental change and conflict in the global south since the Second World War (Goldman 2005; Rich 2013), as we also saw in the case of water privatisation in Chapter 6.

The Global Energy Atlas visualises the potential that exists for converting solar radiation into electricity in a photovoltaic solar panel. It uses a measure known as "yield" (kWh/kWp), which is a standard performance indicator for solar systems. In thermodynamic terms, a high yield translates into high productivity. Starting from Haraway's assertion that there is "no gaze from nowhere", James McCarthy and Jim Thatcher (2019) analyse the political work done by the Global Solar Atlas. The Atlas provides an image of the world in terms of yield, and thereby, they argue, it makes a claim to what the environment represents. This representation allows the

FIGURE 7.2 Map from the Global Solar Atlas showing the potential for photovoltaic solar energy development in South America

A dark red colour (such as that on the Peruvian coast) indicates a high "yield", whereas a yellow or green colour (such as that in coastal Ecuador) indicates a lower yield.

Map obtained from the Global Solar Atlas 2.0, "a free, web-based application developed and operated by the company Solargis s.r.o. on behalf of the World Bank Group, utilizing Solargis data, with funding provided by the Energy Sector Management Assistance Program (ESMAP)". For additional information: globalsolaratlas.info.

World Bank to fund solar electrification projects and pay dividends to its shareholders, but at a deeper level, it also demonstrates how knowledge always is situated in a social and institutional context. In this sense, the Global Solar Atlas and the map used in the process of diamond exploration in Chapter 2 are no different.

Meantime, the map does not recognise the many other meanings that exist in and of the land it depicts: what the Andes and *Pachamama* (Mother Earth) represent in indigenous Aymara and Quechua cosmologies (Burman 2017); how the construction of solar farms would change the gendered dynamics associated with different land uses in the Bolivian valleys (Paulson 2005); how changes in the energy landscape would upset power relations in urban Chile (Castán Broto 2019); or how off-grid solar energy systems would affect the livelihood strategies of villagers in the Amazon forest (Mazzone 2019). Rooted in a Western epistemology that aspires to universal applicability, the energy idea silences "epistemologies of the South" (Santos 2014). The map itself is a powerful tool that naturalises a situated, partial perspective, and as McCarthy and Thatcher (2019: 245, original emphases) conclude, the Global Solar Atlas first and foremost serves power:

> [T]he maps … both make territory *more* legible, increasing what states and capitals can "see" in certain respects (i.e. detailed quantitative data regarding solar or wind potential, including locations, intensity, and potential values), while making other things *less* legible, such as existing land uses and claims omitted from many of the maps. They thus depict the territories they represent as empty and available, yet full of resources and potential for investors.

EMBEDDED ENERGY

Not long ago, one of us found himself in a residential area known as Ciudad Nuclear (Nuclear City) in southern Cuba (Figure 7.3). Upon entering the area, the first building on the right was adorned with the image of a clenched fist emerging from an electric flash. The second building had no walls, but in the third, life was going on just as normal. The clenched fist signalled the presence of the Cuban state-owned electrical utility, Unión Eléctrica, and the local utility branch was named the CEN—the abbreviated form of *Central Electronuclear*—Spanish for Nuclear Power Plant. Powerlines crossed overhead. The CEN delivered electricity to light up the state-run ration shop and the local school, and it brought power to the ubiquitous electric pressure cookers in Ciudad Nuclear's apartment blocks. To use a concept from Chapter 5, it was central to sustaining the urban metabolism. While Ciudad Nuclear in many respects was a place like any other, there was also something perplexing about it: there is no nuclear power plant in Cuba.

The energy idea, as we have seen, is a form of situated knowledge, but it nevertheless gives us a conceptual toolkit with which we can make sense of a place like Ciudad Nuclear. In ecological terms, the laws of thermodynamics are fundamental to explaining the "circulations and metabolisms" that give rise to urban environments—cities, like economic systems more broadly, need a continuous supply of energy and other natural resources or else human activities will come to a stop. Thus, the laws of thermodynamics also help to explain why societies become dependent on particular energy forms. If energy infrastructures provide low entropy to human activities, such as the use of pressure cookers in Ciudad Nuclear, energy cannot be seen as incidental to socio-economic life. Rather, different energy forms become embedded in human social and economic activities and their rhythms (Walker 2021). Ciudad Nuclear's peculiar landscape, therefore, allows us to see how energy systems and socio-economic systems develop together. They are in a co-determining relationship, part of an integrated political ecological process.

FIGURE 7.3 Housing and powerlines in Ciudad Nuclear, Cuba

Photo by Gustav Cederlöf

At the time of the Cuban Revolution, Fidel Castro and other political leaders argued that the Cuban economy had developed for one sole purpose: sugar production. During Spanish colonial rule and later US neo-colonial dependence, the Cuban majority population toiled in the cane fields, but the sugar and the profits it generated ended up in the metropoles. Echoing arguments around "dependent urbanisation", which suggested that cities in Latin America had developed for different political economic reasons than cities in Europe (Chapter 3), the Cuban revolutionaries argued that Havana had developed into a "parasitical city". Its purpose was to administer the extractive colonial economy rather than to serve as a centre for domestic industrialisation and development. In the capital, the white elite enjoyed access to electricity and other modern amenities, but in the *barrios* and the city's rural hinterlands, the racialised working population remained in poverty. Based on this analysis, the revolutionary government decided to neglect Havana and instead set out to bring socialist modernity to the entire population.

Gustav Cederlöf (2020) demonstrates how the resulting urban programme was intimately connected with a kind of energy infrastructure. The Cuban government reasoned that electricity, materially and culturally, was an enabling condition for modern urban living. Electricity would power refrigerators and TVs, hospital equipment and the cinema. However, the significance of centralised electricity infrastructure extended beyond the practices it enabled. If the infrastructures of the socialist state reached into every household and every workplace, the government could also "rationally" distribute energy to the population, not as a commodity but as an entitlement. Networked electricity infrastructure was, therefore, seen to provide the conditions of possibility for "energy justice" (Jenkins et al. 2016; Hornborg et al. 2019) and for the socialist state as a form of socio-ecological organisation (Cederlöf 2023).

The urban-infrastructural project also had an international dimension. To generate electricity, the Cuban socialist state relied on oil, which it imported from the Soviet Union. Nevertheless, the government envisioned that this fossil fuel-dependent infrastructure eventually would give way to the most advanced "productive forces" available in the world—nuclear energy—which would reduce Cuba's dependence on international oil trade. In 1983, work also started on a first nuclear reactor outside the city of Cienfuegos, and next to it, a new residential area was planned to house the nuclear workforce in the most modern form of urban development. Ciudad Nuclear was a manifestation of "nuclear modernity" and epitomised the promises of revolutionary change in the global south.

Through a close study of Ciudad Nuclear, it is evident how electricity, oil and nuclear energy became embedded in the Cuban socialist project and in the environment they produced for human life. They formed part of a political ecology in the ontological sense of the term. The three energy sources were entrenched in the physical landscape but also in the political economic logic of the socialist state and in the cultural significance of Cuba as a modern, socialist country. Bridge et al. (2013: 333) argue that energy systems in this way become embedded in particular spatial configurations: "the networked nature of the system itself produces geographies of connection, dependency and control", they write. In the early 1990s, no Cuban could fail to notice how fossil energy had become an enabling condition of everyday life. When the Soviet Union collapsed, Cuba lost support for its nuclear project, which left a half-finished nuclear reactor and an only partially built Ciudad Nuclear on the island. But more drastically, the country also lost more than 85 per cent of its oil supplies, leading to daily blackouts, paralysing transports and crippling the economy. In the time since, Cubans have been forced to adapt to living with oil scarcities, prompting them to reconsider notions of socialist urban development and produce new spatial patterns of economic activity (Cederlöf 2023).

Ciudad Nuclear can be seen as a case study that illustrates how particular energy forms become embedded in the spatial patterns of social and economic activity. Other recent work by political ecologists has continued to elaborate on this theme. For example, Maputo is the capital of Mozambique, and Vanesa Castán Broto's (2019) research on energy use in the city is a case in point. The Portuguese established Maputo in the 1780s as a colonial hub that linked roads from sites of extraction in Mozambique's resource-rich interior with the sea-based export market. Until Mozambique's independence in 1975, the city was called Lourenço Marques after a sixteenth-century Portuguese coloniser. Maputo developed along racialised lines: around the port, the white "cement city" (*cidade de cimento*) was administratively and infrastructurally separated from the surrounding "reed city" (*cidade de caniço*), which housed black Mozambican labourers. Today, the areas intersect, but while electricity and liquefied petroleum gas (LPG) are readily available in the cement city, charcoal is the dominant energy source in the *bairros* that subdivide the traditional reed city.

Walking the *bairros*, Castán Broto (2019) notes how particular forms of energy are embedded in different social practices. For instance, electricity is generally reserved for communications, such as mobile phone charging, whereas charcoal is used for cooking and water heating. Idalina Baptista (2016) argues that the Maputo residents tend to favour prepayment arrangements to access electricity since this increases their autonomy vis-à-vis the national utility. Castán Broto similarly argues that charcoal has become embedded in household activities, such as cooking, since it increases the households' self-determination. Thus, the use of charcoal is politically motivated. Charcoal enters the urban economy from Maputo's hinterlands and is progressively split into smaller quantities by wholesalers and distributors along the road. The urban residents are then able to purchase charcoal on the street corner to cover their daily needs for cooking, allowing them to monitor and control what they consume. "The charcoal network constitutes an alternative to obscure systems of distribution for electricity and petrol", Castán Broto (2017: 241) explains. "When it comes to charcoal, local business people control the whole chain". Hence, charcoal is also embedded in a spatially manifest economic system that involves a multitude of small businesses and that links urban with rural livelihoods, testifying to a process of extended urbanisation (Chapter 5).

If we compare Ciudad Nuclear and Maputo with another urban setting, we can appreciate why oil has become such an important energy source in North American everyday life. Figure 7.4 is a satellite image showing a part of the city of Columbus, Ohio. Each light-coloured mark along the grid-like road network is a driveway leading up to a family home. Suburban development like this took off on a large scale in the United States in the 1930s, and as Matthew Huber (2013) argues, few things became more integral to a person's pursuit of the "American way of life" than home ownership—"specifically", he writes, "a private, detached, single-family home in a suburban location" (38). The upsurge in suburban property ownership occurred alongside a rapid increase in car registrations. A car, expressing one's identity and suiting one's needs, enabled a growing US middle class to commute from home to work, to school and the shopping mall. Cars and sprawling infrastructural networks of petrol supplies thus provided the enabling conditions for a privatised form of mobility, which was enmeshed in a cultural and economic project. Car ownership, Huber argues, "allowed for a privatized command over and experience

FIGURE 7.4 Satellite imagery over a suburb of Columbus, Ohio

Sources: CNES/Airbus, Mazar Technologies, U.S. Geological Survey, USDA/FPAC/GEO/Google Maps 2022

of space that became constitutive of American conceptions of freedom, the open road, and the crafting of … 'autonomous subjects'" (74). Via the automotive power of the car engine, oil was embedded in the built environment of cities like Columbus and in the spatial patterns of the neoliberal US economy more generally. As a result, it is difficult to separate the experience of using oil from the cultural significance of being an "American" in suburban United States.

BETWEEN DEGROWTH AND GREEN GROWTH

The embedded perspective on energy provides an entry point for political ecologists to examine the conditions of a low-carbon energy transition. Clearly, an energy transition has thermodynamic and technological implications, but it is also firmly connected to issues of socio-economic, political and cultural change. Using an integrative, interdisciplinary approach that is sensitive to all of these aspects, many political ecologists align themselves with a radical view of energy transitions that seeks to transform the economic and cultural patterns within which fossil fuels are embedded. The case for "degrowth" starts with the observation that an exponentially growing economy requires ever-increasing supplies of low-entropy energy and raw materials. The degrowth hypothesis holds that such a cornucopian economic logic ultimately is impossible to sustain within the limits of a finite planet. Eventually, the engine runs out of fuel. At the very least, exponential growth will generate increasingly violent political conflicts associated with energy and mineral extraction (D'Alisa et al. 2014; Paulson 2017; Kallis et al. 2020).

The degrowth approach to sustainability rests on three core arguments. First, it is necessary to reduce the levels of energy and material throughput in the global economy in thermodynamic terms—making the social metabolism smaller—since climate change and environmental degradation result from excessive consumption. Second, economic life has to be organised around notions of well-being that transcend the pursuit of growth. Echoing an understanding of political ecology as praxis—i.e. that we create alternative worlds by *doing* things differently—an important degrowth strategy is to engage with actually existing experiments of using community currencies (Hornborg 2019), universal basic income (Demaria et al. 2019) and projects that "keep fossil fuels in the ground" (Princen et al. 2015). Degrowth activists also work with indigenous communities who traditionally organise social and economic life around conceptions other than the energy idea and the growth imperative (Kothari et al. 2019). Third, these existing alternatives show that a voluntary downscaling of consumption and production in the global north is both possible and desirable to achieve.

In contrast, many other political ecologists concern themselves with more mainstream approaches to energy transitions. The idea of "green growth" dominates national and international agendas and is encapsulated in the UN's 17 Sustainable Development Goals (sdgs.un.org). It builds on the conviction that the mobilities and the consumption practices of the wealthier parts of the global population are desirable and therefore need to be maintained (Adams 2020). Advocates of green growth promote large-scale investments in low-carbon technologies and increasing energy efficiency to reduce the demand for fossil fuels while continuously increasing the size of the economy. While green growth is promoted by some as a neoliberal solution to the climate emergency (Bakker 2015), it is also the key driver for change in Keynesian-inspired proposals, such as eco-modernist approaches and the so-called Green New Deal (Aronoff et al. 2019; Knuth et al. 2020). Just as degrowth, green growth is a political choice: by appealing to existing consumption practices with technological innovations, such as electric vehicles (Urry 2016), smart cities (Marvin et al. 2016) or industrial methods of sequestering carbon (Carton et al. 2020), it aims to sustain (or in the case of a Green New Deal reform) the economic interests and power relations manifest in the landscapes historically enabled by fossil fuels.

Unavoidably, green growth generates rapid demand for renewable energy and raw materials to produce new technologies. This is notably evident in the case of lithium-ion batteries—a relatively new kind of battery at the epicentre of efforts of decarbonisation. Used in electric cars and bikes, mobile phones and laptops, but also entire electricity grids incorporating renewable energy sources, lithium has become the subject of rampant economic speculation, booming production and mounting political conflict. Almost 60 per cent of the world's reserves of this light metal are estimated to exist in the salt flats that cut across the borders of Argentina, Bolivia and Chile. Here, lithium production takes place in landscapes inhabited by indigenous populations, marked by centuries of colonialism and extractivism (Bustos-Gallardo et al. 2021; Kingsbury 2022). Thus, as the case of lithium demonstrates, green growth turns a low-carbon energy transition into a question of how, where and by whom energy and materials are *extracted* and how the costs and benefits of such extraction are distributed locally and globally.

EXTRACTED ENERGY

To think about energy as something extracted is to give it the conceptual form of a resource: an object harnessed from nature for human benefit. In contrast to the embedded perspective, which develops a relational approach to energy, energy extraction establishes a distinction between the "human" and the "natural" as two separate, albeit interdependent domains (Cederlöf 2021). Political ecologists trained in anthropology have provided some of the most detailed and disturbing accounts of how mass consumption in certain places generates environmental conflicts in sites of extraction, ranging from the militarised landscapes of oil and coal production in Northeast India (Kikon 2019) to the settler-colonial legacies of coal and uranium mining in the Navajo Nation (Powell 2018) and controversial wind energy development in Mexico (Howe 2019). The extractive process can be pictured in terms of a dual movement: first, it delivers low-entropy energy and materials from natural stocks (such as oil wells and lithium deposits) or natural flows (such as wind and water) to economic activities, and in these activities, energy and materials are degraded while economic value is produced (Georgescu-Roegen 1971). Second, fossil fuel extraction delivers carbon from the subsoil to the atmosphere at a greater rate than that by which carbon is captured by forests, oceans, soils and human technological innovations, contributing to climate change (Foster 1999; Clark and York 2005).

Political ecologists have explored the extractive process by developing analyses sensitive to power (Chapter 2) and scale (Chapter 3), and one additional concept that helps to examine how a resource links local experiences with global processes is the "commodity chain". A commodity chain captures how a consumer is but one actor in a greater political economic context of energy, raw materials and waste flows. In the case of oil, the commodity chains we become part of at the point of consumption are truly global: "To fill up at the local gas station", as Bridge and Philippe Le Billon (2017: 40) argue, "is also to become entangled in concerns about national security, drill rig safety, Arctic sea ice, financial speculation, and indigenous rights". Figure 7.5 shows a generally outlined commodity chain for oil products. However, it is possible to trace the commodity chain of any energy resource—fossil or renewable—and

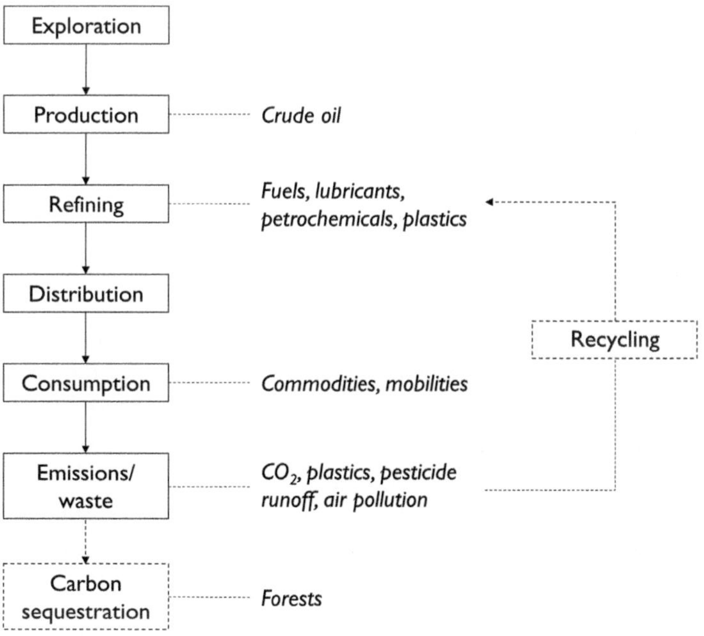

FIGURE 7.5 Generally outlined commodity chain for oil products

it may be globally distributed, such as in the cases of biofuels (McGrath 2013), wind turbines (Baker and Sovacool 2017) and photovoltaic solar panels (Mulvaney 2019), or it may be more local, such as in the case of Mozambican charcoal.

Exploration

Oil extraction starts with exploration. This is an activity in which states and firms survey where oil might be found using geophysical methods. Exploration is a deeply contested part of the extractive process in which politically and economically powerful actors map the world in terms of resource potential and make claims to land for production. It is important to take the materiality of the resource searched for into consideration in order to critically analyse the exploration process (Bakker and Bridge 2021). In contrast to most renewable energy sources, fossil fuels have a point-like, discontinuous distribution in space. Wells and mines are vertical subterranean spaces which are tapped from above. This "geography of holes" prompts oil companies to carve out an extractive enclave from its local surroundings and integrate it into the cultural, legal and infrastructural context of the globalised oil industry (Bridge 2011; Watts 2012).

In a study of oil extraction in Canada's tar sands, Anna Zalik (2015) demonstrates how transnational oil firms frequently initiate formal reviews of potential oil-field developments. In doing so, they also establish the regulatory framework within which other actors, such as First Nations, later have to act in order to contest extractive activities. Indeed, the extraction of oil from bitumen in the province of Alberta is a window on the multifaceted conflicts and the legal frameworks within which they play out that emerge in sites of energy extraction. These conflicts involve national interests, big oil, the destruction of forest and wetland habitats, workers' rights, identity as well as the survival and livelihoods of indigenous communities (Adkin 2016). Just as the environmental knowledge presented in the Global Solar Atlas and the map encountered in case of diamond mining in Chapter 2, oil exploration creates expert knowledge that legitimises extraction and serves the industry's economic interests. Simultaneously, it marginalises local, often non-Western knowledges and lived experiences of the landscape.

Production

The second link in the commodity chain is production. Here, states and firms are again the most powerful actors. States often concede land to oil companies in return for rent, while oil companies invest capital to turn a profit. Production, therefore, generates spaces of intense value creation, and these are also spaces in which capital, states and the citizens they represent negotiate how that value should be distributed (Bridge and Le Billon 2017; Purcell and Martinez 2018). Political ecologists have raised several questions in this context: for whom is value created? Who is exposed to the risks of production? And what happens to the site of production once extraction ends? The notion of dispossession is key to exposing the politics of production, because one actor's access to a drilling site tends to imply that another actor is dispossessed of that land and resource wealth (Sawyer 2004; Watts 2012). On a national scale, as we discuss at length in Chapter 8, dispossession often also occurs through a process of enclosure, when public assets or common lands are turned into private property for energy development (Baka 2017).

In a fascinating study, Hannah Appel (2012) takes the reader to an oil extraction enclave in Equatorial Guinea. The enclave's residents are oil workers who have flown in from Texas, and they compare their African home to a suburb of Houston. But Appel contrasts the material living conditions within the enclave with life on its outside and demonstrates how the extractive process produces stark racialised and class-based inequalities in the environment—a kind of "infrastructural violence". She writes:

> Enclave construction not only forces dispossession on many local residents, but also symbolizes the lack of access to services such as potable water, reliable electricity, and other forms of domestic comfort in Equatorial Guinea beyond the walls. The enclaves' private luxury, in other words, stands as a securitized monument to escalating petro-inequality.
> (441–442)

Refining

When oil is extracted, a thick, nasty substance known as crude oil comes out of the ground. For crude oil to be useful, it has to be refined. This is the third link in the commodity chain. A refinery works like a distillery: you heat up the crude oil and at quite low temperatures the so-called light fractions evaporate. These are petroleum gases, which can be used as a cooking fuel. At higher temperatures, kerosene and petrol are separated out, before the heavier fractions such as diesel and fuel oil are distilled. Refineries also produce feedstock, which is used to make plastics and as a resource within the refinery itself to produce other petrochemicals. Huber (2013) argues that refineries are often invisible but entirely central nodes in the political ecological relations that produce and reproduce the neoliberal US economy, along with the experience of "being an American". Without refineries, there would be no petrol available to sustain the spatial patterns of suburban mobility in a place like Columbus. Meantime, refineries have a hefty environmental impact. Leaks and spills are common, waste greenhouse gases are often flared into the atmosphere, and carcinogens such as benzenes are inevitable by-products. As Janice Harper (2004: 308–309) finds in one pioneering study, the highest levels of air pollution in the US "energy capital" of Houston were concentrated in the parts of the city where poor and minority communities lived, "closer to the Houston ship channel, oil refineries, petrochemical industries, and other sources of air pollution". Air pollution from refining thus intersected with the patterns and lived experiences of social inequality.

Distribution

Between sites of production, refining and consumption, oil and its derivatives need to be distributed. It is difficult to overemphasise the importance of energy transport: if movement stops in pipelines, powerlines and at sea, the global economy grinds to a halt. When Soviet oil shipments to Cuba stopped in the early 1990s, the Cuban economy diminished by over one-third in size, in large part resulting from the lack of low-entropy energy in the metabolic process (Cederlöf 2023). Reflecting the threat of socially produced energy scarcity such as this, energy transports are a vital concern in national and international politics, and disrupted energy transports is often seen as a geopolitical "weapon". While studies in this area traditionally have focused on the geopolitics of oil and gas (Yergin 2008; Bridge and Le Billon 2017), scholars have recently started investigating how renewables reconfigure and give rise to new geopolitical relations (Overland 2019; Fischhendler et al. 2022).

Historically, energy transports have also played a more subtle political role than to act as a geopolitical weapon. In *Carbon Democracy*, Timothy Mitchell (2011) demonstrates how coal became the all-important energy source in the British economy during the Industrial Revolution. Coal was cut in bulk and then transported on railways for use in places like the cotton mills of Glasgow. The distinct material characteristics of coal meant that organised workers could attain a particular kind of political power: by going on strike at coalfaces and slowing down transports on the railway—impeding the economic process with reduced low-entropy supplies—they could make political demands for increasing democratic influence. Mitchell argues that the transition from coal to oil in Europe, therefore, above all was politically motivated. The fluidity of oil, as opposed to the solidity of coal, meant that energy could be transported in closed pipelines across land and in tankers at sea with relatively little labour effort. This made the work of coal heavers and stokers redundant. Most oil-producing states, Mitchell observes, are also authoritarian states, and while careful to avoid determinist explanations, he argues that this political outcome has to be explained in relation to European and North American neo-colonial efforts of securing continuous energy supplies in the Middle East. On Barak (2020) has recently provided an important corrective to Mitchell's argument. Coal not only powered the British economy, he points out, but also British imperialism—or "coalonialism" as he calls it. Alongside oil, coal has been intrinsic to a much longer history of autocratic government in the Middle East, notably as it metabolised a vast transport infrastructure interconnecting the British Empire in the region and beyond.

Consumption

Next in the commodity chain is consumption. This is where the landscapes in which particular energy sources are embedded enter the extractive process. To complicate things, it is important to note that a landscape of energy production at the same time is a landscape of energy consumption. The notion of "energy return on investment" (EROI) is useful to conceptualise this point. Put simply, an extractive process must generate more low entropy than it dissipates or else there will be a net loss of energy. Charles Hall et al. (2009) find that a ratio of 3:1 is the minimum EROI necessary for oil and liquid biofuels to reach an end-consumer in the United States. This means that for each barrel of oil brought to the market, three have to be produced to account for the energy expended in production, refining and transport. Traditionally, fossil fuel extraction has had a very high EROI compared to renewable energies. For instance, Daniel Weiβbach et al. (2013) conclude that coal extraction has an EROI of 30:1 while German photovoltaic solar panels fall short of 4:1. The question, then, is if renewable energy technologies have a sufficiently high EROI to sustain a high-technological society without themselves having to be subsidised by fossil energy sources. One recent study found that if a complete renewable energy transition can be achieved in electricity production by 2060, the EROI of the global energy system would decrease from about 12:1 to between 3:1 and 5:1 (Capellán-Pérez et al. 2019). In any case, the consumption of energy in landscapes of production indicates that the "commodity chain" is a simplified, linear model of a complex process that folds back on to itself at several points in terms of production, refining, transport and consumption.

Emissions and waste

The final link in the commodity chain is emissions and waste. In the case of oil, these result from combustion and the use of plastics and petrochemicals. While CO_2 emissions from fossil fuel combustion reached a high in 2018, the emissions of greenhouse gases decreased sharply during the COVID-19 pandemic. The close correlation between economic growth and emissions is evident from the environmental impact of the recession that followed in the wake of worldwide lockdowns. The overall greenhouse gas emissions from energy then saw the largest year-on-year increase ever as the world opened up in 2021 to exceed pre-pandemic levels (IEA 2022). Meanwhile, the infamous Great Pacific Garbage Patch was recently estimated to contain 79,000 tonnes of ocean plastic (Lebreton et al. 2018), and microplastics (plastics less than 5 mm in size) are now ubiquitous in the marine environment. Microplastics notably derive from car tyre wear, the washing of synthetic textiles and the use of cosmetics, and they have a strong capacity of adsorbing environmental pollutants. Thus, when microplastics are ingested by marine organisms, they can cause immediate physical damage to the organism through cuts but also more long-term harm when bioaccumulated toxins are transferred up the food chain (Wang et al. 2016).

On a more positive note, plastics and petrochemicals are increasingly recycled in efforts to create a more "circular economy" that reduces the demand for newly extracted resources. Nevertheless, the Second law of thermodynamics asserts that complete circularity is impossible to achieve in a world of increasing entropy (D'Alisa 2019). Carbon emissions are also sequestered in carbon sinks, such as forests and oceans. Recent assessments by the International Panel on Climate Change (IPCC) indicate that enhanced carbon sequestration—what sometimes is referred to as "negative emissions"—represents one of the greatest opportunities but also challenges for addressing climate change (Carton et al. 2020). While CO_2 increasingly is captured at the point of combustion to be stored in underground spaces—again a highly energy intensive activity—forests are the most important carbon sinks globally. The fossil fuel commodity chain, therefore, can be seen to end not only in the atmosphere but also in the forest, and in recent years, forests have become one of the most contested spaces in national and international efforts of climate adaptation. This is a subject we return to in the next chapter.

CONCLUSION AND LEARNING OUTCOMES

Climate change necessarily brings attention to energy and its politics. The concept of "energy" itself has a relatively short history and developed during the Industrial Revolution, where it reflected the situated perspective of engineers and factory owners in the core of an imperial economy. When the environment is spoken of in terms of "energy", the world is ruled by an economistic logic set against the existential threat of scarcity. While green growth operates within this logic to achieve a low-carbon transition—generating extractive conflicts—advocates of degrowth engage with existing alternative ways of perceiving the environment and show that it is possible to do things differently.

In keeping with the energy idea, energy can be understood as something *embedded* in human activities and the spatial patterns that enable these. The embedded perspective allows us to analyse how particular energy forms have been co-productive of social relations of production over time. We can also understand energy as something *extracted*, delivering work potential and carbon from the subsoil via landscapes of consumption to the atmosphere. Using the concept of the commodity chain, it is possible to see that the extractive process throughout is linked to the exercise of social power in four main ways:

1. *Dispossession*. While some actors gain access to resources through exploration and production, the extractive process tends to deprive other actors of land, property and resource wealth.
2. *Marginalisation*. While energy extraction benefits some economic interests and social groups in landscapes of consumption and production, it tends to marginalise the livelihoods, knowledge and culture of communities living on the resource frontier.
3. *Degradation*. While extraction generates economic value, it also degrades the environment as entropy increases. The costs and benefits of leaks, spills, emissions and waste are spread unevenly among social groups and nonhuman species.
4. *Risk*. While consumers tend to benefit from resource use, the risks of extraction are displaced to extractive environments and adjacent areas inhabited by local communities.

Having read this chapter, we hope that you can:

- Account for the origins of the energy idea as a form of situated Western knowledge
- Explain how the energy idea serves power and is linked to a history of colonisation
- Assess how particular energy forms are embedded in landscapes of consumption, where they serve social, economic, cultural and political interests
- Analyse how energy extraction takes place along a commodity chain, linking environmental actors in an often-global network with socially and economically unequal outcomes

FURTHER READING

Daggett, Cara New. 2019. *The Birth of Energy: Fossil Fuels, Thermodynamics, and the Politics of Work*. Durham, NC: Duke University Press.

The Birth of Energy is a rich yet accessible text to read after this chapter if you want to explore the implications of a thermodynamic understanding of energy and the epistemological politics associated with the energy idea.

Kikon, Dolly. 2019. *Living with Coal and Oil: Resource Politics and Militarization in Northeast India*. Seattle, WA: University of Washington Press.

Kikon's work is a fantastic example of political ecological scholarship in a violent environment that offers insights on themes of extraction, meaning, and marginalisation on the resource frontier. It is a gripping ethnographic account of life in an extractive energy landscape.

BIBLIOGRAPHY

Adams, W. M. 2020. *Green Development: Environment and Sustainability in a Developing World*, 4th ed. London, UK: Routledge.

Adkin, Laurie E., ed. 2016. *First World Petro-Politics: The Political Ecology and Governance of Alberta*. Toronto, ON: University of Toronto Press.

Appel, Hannah. 2012. "Walls and White Elephants: Oil Extraction, Responsibility, and Infrastructural Violence in Equatorial Guinea." *Ethnography* 13(4): 439–465.

Aronoff, Kate, Alyssa Battistoni, Daniel Aldana Cohen, and Thea Riofrancos. 2019. *A Planet to Win: Why We Need a Green New Deal*. London, UK: Verso.

Baka, Jennifer. 2017. "Making Space for Energy: Wasteland Development, Enclosures, and Energy Dispossessions." *Antipode: A Radical Journal of Geography* 49(4): 977–996.

Baker, Lucy, and Benjamin K. Sovacool. 2017. "The Political Economy of Technological Capabilities and Global Production Networks in South Africa's Wind and Solar Photovoltaic (PV) Industries." *Political Geography* 60: 1–12.

Bakker, Karen. 2015. "Neoliberalization of Nature." In *Routledge Handbook of Political Ecology*, edited by Tom Perrault, Gavin Bridge, and James McCarthy, 446–456. London, UK: Routledge.

Bakker, Karen, and Gavin Bridge. 2021. "Material Worlds Redux: Mobilizing Materiality within Critical Resource Geography." In *Routledge Handbook of Critical Resource Geography*, edited by Matthew Himley, Elizabeth Havice, and Gabriela Valdivia, 43–56. London, UK: Routledge.

Baptista, Idalina. 2016. "Maputo: Fluid Flows of Power and Electricity—Prepayment as Mediator of State-Society Relationships." In *Energy, Power and Protest on the Urban Grid: Geographies of the Electric City*, edited by Andrés Luque-Ayala and Jonathan Silver, 112–131. London, UK: Routledge.

Barak, On. 2020. *Powering Empire: How Coal Made the Middle East and Sparked Global Carbonization*. Oakland, CA: University of California Press.

Bridge, Gavin. 2011. "Past Peak Oil: Political Economy of Energy Crises." In *Global Political Ecology*, edited by Richard Peet, Paul Robbins, and Michael J. Watts, 307–324. London, UK: Routledge.

Bridge, Gavin, Stefan Bouzarovski, Michael Bradshaw, and Nick Eyre. 2013. "Geographies of Energy Transition: Space, Place and the Low-Carbon Economy." *Energy Policy* 53: 331–340.

Bridge, Gavin, and Philippe Le Billon. 2017. *Oil*, 2nd ed. Cambridge, UK: Polity Press.

Burman, Anders. 2017. "The Political Ontology of Climate Change: Moral Meteorology, Climate Justice, and the Coloniality of Reality in the Bolivian Andes." *Journal of Political Ecology* 24: 921–938.

Bustos-Gallardo, Beatriz, Gavin Bridge, and Manuel Prieto. 2021. "Harvesting Lithium: Water, Brine and the Industrial Dynamics of Production in the Salar de Atacama." *Geoforum* 119: 177–189.

Capellán-Pérez, Iñigo, Carlos de Castro, and Luis Javier Miguel González. 2019. "Dynamic Energy Return on Energy Investment (EROI) and Material Requirements in Scenarios of Global Transition to Renewable Energies." *Energy Strategy Reviews* 26: 100399.

Carton, Wim, Adeniyi Asiyanbi, Silke Beck, Holly J. Buck, and Jens F. Lund. 2020. "Negative Emissions and the Long History of Carbon Removal." *WIREs Climate Change* 11(6): 1–25.

Castán Broto, Vanesa. 2017. "Energy Sovereignty and Development Planning: The Case of Maputo, Mozambique." *International Development Planning Review* 39(3): 229–248.

Castán Broto, Vanesa. 2019. *Urban Energy Landscapes*. Cambridge, UK: Cambridge University Press.

Cederlöf, Gustav. 2020. "The Revolutionary City: Socialist Urbanisation and Nuclear Modernity in Cienfuegos, Cuba." *Journal of Latin American Studies* 52(1): 53–76.

Cederlöf, Gustav. 2021. "Out of Steam: Energy, Materiality, and Political Ecology." *Progress in Human Geography* 45(1): 70–87.

Cederlöf, Gustav. 2023. *The Low-Carbon Contradiction: Energy Transition, Geopolitics, and the Infrastructural State in Cuba*. Oakland, CA: University of California Press.

Clark, Brett, and Richard York. 2005. "Carbon Metabolism: Global Capitalism, Climate Change, and the Biospheric Rift." *Theory & Society* 34: 391–428.

Coppola, Damon P. 2021. *Introduction to International Disaster Management*, 4th ed. Amsterdam, Netherlands: Butterworth-Heinemann.

D'Alisa, Giacomo. 2019. "Circular Economy." In *Pluriverse: A Post-Development Dictionary*, edited by Ashish Kothari, Ariel Salleh, Arturo Escobar, Federico Demaria, and Alberto Acosta, 28–30. New Delhi, India: Tulika Books.

D'Alisa, Giacomo, Federico Demaria, and Giorgos Kallis. 2014. *Degrowth: A Vocabulary for a New Era*. London, UK: Routledge.

Daggett, Cara New. 2019. *The Birth of Energy: Fossil Fuels, Thermodynamics, and the Politics of Work*. Durham, NC: Duke University Press.

Demaria, Federico, Giorgos Kallis, and Karen Bakker. 2019. "Geographies of Degrowth: Nowtopias, Resurgences and the Decolonization of Imaginaries and Places." *Environment and Planning E: Nature and Space* 2(3): 431–450.

Feynman, Richard. 2013 (1963). *The Feynman Lectures on Physics*, vol. 1. New York, NY: Basic Books.

Fischhendler, Itay, Lior Herman, and Lioz David. 2022. "Light at the End of the Panel: The Gaza Strip and the Interplay between Geopolitical Conflict and Renewable Energy Transition." *New Political Economy* 27(1): 1–18.

Forsyth, Tim. 2015. "Integrating Science and Politics in Political Ecology." In *International Handbook of Political Ecology*, edited by Raymond L. Bryant, 103–116. Cheltenham, UK: Edward Elgar.

Foster, John Bellamy. 1999. "Marx's Theory of Metabolic Rift: Classical Foundations for Environmental Sociology." *American Journal of Sociology* 105(2): 366–405.

Fridays for Future. 2022. "Strike Statistics." Fridays for Future website, https://fridaysforfuture.org/what-we-do/strike-statistics. Accessed: 28 August 2022.

Georgescu-Roegen, Nicholas. 1971. *The Entropy Law and the Economic Process*. Cambridge, MA: Harvard University Press.

Goldman, Michael. 2005. *Imperial Nature: The World Bank and Struggles for Social Justice in the Age of Globalization*. New Haven, CT: Yale University Press.

GSA. 2021. "About." Global Solar Atlas website, https://globalsolaratlas.info/support/about. Accessed: 16 April 2021.

Hall, Charles A. S., Stephen Balogh, and David J. R. Murphy. 2009. "What Is the Minimum EROI that a Sustainable Society Must Have?" *Energies* 2: 25–47.

Haraway, Donna. 1991. *Simians, Cyborgs, and Women: The Reinvention of Nature*. London, UK: Free Association Books.

Harper, Janice. 2004. "Breathless in Houston: A Political Ecology of Health Approach to Understanding Environmental Health Concerns." *Medical Anthropology* 23: 295–326.

Hornborg, Alf. 2019. *Nature, Society, and Justice in the Anthropocene: Unraveling the Money-Energy-Technology Complex*. Cambridge, UK: Cambridge University Press.

Hornborg, Alf, Gustav Cederlöf, and Andreas Roos. 2019. "Has Cuba Exposed the Myth of 'Free' Solar Power? Energy, Space, and Justice." *Environment and Planning E: Nature and Space* 2(4): 989–1008.

Howe, Cymene. 2019. *Ecologics: Wind and Power in the Anthropocene*. Durham, NC: Duke University Press.

Huber, Matthew T. 2013. *Lifeblood: Oil, Freedom, and the Forces of Capital*. Minneapolis, MN: University of Minnesota Press.

IEA. 2020. *CO_2 Emissions from Fuel Combustion: Overview*. Paris, France: International Energy Agency.

IEA. 2022. *Global Energy Review: CO_2 Emissions in 2021*. Paris, France: International Energy Agency.

Illich, Ivan. 2010 (1983). "The Social Construction of Energy." *New Geographies* 2: 11–19.

Jenkins, Kirsten, Darren McCauley, Raphael Heffron, Hannes Stephan, and Robert Rehner. 2016. "Energy Justice: A Conceptual Review." *Energy Research & Social Science* 11: 174–182.

Jørgensen, Sven Erik. 2012. *Introduction to Systems Ecology*. Boca Raton, FL: CRC Press.

Kallis, Giorgos, Susan Paulson, Giacomo D'Alisa, and Federico Demaria. 2020. *The Case for Degrowth*. Cambridge, UK: Polity Press.

Kikon, Dolly. 2019. *Living with Coal and Oil: Resource Politics and Militarization in Northeast India*. Seattle, WA: University of Washington Press.

Kingsbury, Donald V. 2022. "Lithium's Buzz: Extractivism between Booms in Bolivia, Argentina, and Chile." *Cultural Studies*. DOI: 10.1080/09502386.2022.2034909.

Knuth, Sarah, John Stehlin, and Nate Millington. 2020. "Rethinking Climate Futures through Urban Fabrics: (De)growth, Densification, and the Politics of Scale." *Urban Geography* 41(10): 1335–1343.

Kondepudi, Dilip, and Ilya Prigogine. 1998. *Modern Thermodynamics: From Heat Engines to Dissipative Structures*. Chichester, UK: John Wiley.

Kothari, Ashish, Ariel Salleh, Arturo Escobar, Federico Demaria, and Alberto Acosta, eds. 2019. *Pluriverse: A Post-Development Dictionary*. New Delhi, India: Tulika Books.

Lebreton, L., et al. 2018. "Evidence that the Great Pacific Garbage Patch Is Rapidly Accumulating Plastic." *Scientific Reports* 8(4666): 1–15.

Lohmann, Larry. 2016. "What Is the 'Green' in 'Green Growth'?" In *Green Growth: Ideology, Political Economy and the Alternatives*, edited by Gareth Dale, Manu V. Mathai, and Jose A. Puppim de Oliveira, 42–71. London, UK: Zed.

Marvin, Simon, Andrés Luque-Ayala, and Colin McFarlane. 2016. *Smart Urbanism: Utopian Vision or False Dawn?* London, UK: Routledge.

Mazzone, Antonella. 2019. "Decentralised Energy Systems and Sustainable Livelihoods, What Are the Links? Evidence from Two Isolated Villages of the Brazilian Amazon." *Energy and Buildings* 186: 138–146.

McCarthy, James, and Jim Thatcher. 2019. "Visualizing New Political Ecologies: A Critical Data Studies Analysis of the World Bank's Renewable Energy Resource Mapping Initiative." *Geoforum* 102: 242–254.

McGrath, Siobhán. 2013. "Fuelling Global Production Networks with *Slave Labour*? Migrant Sugar Cane Workers in the Brazilian Ethanol GPN." *Geoforum* 44: 32–43.

McNeill, J. R., and Peter Engelke. 2016. *The Great Acceleration: An Environmental History of the Anthropocene since 1945*. Cambridge, MA: Belknap Press of Harvard University Press.

Mitchell, Timothy. 2011. *Carbon Democracy: Political Power in the Age of Oil*. London, UK: Verso.

Mulvaney, Dustin. 2019. *Solar Power: Innovation, Sustainability, and Environmental Justice*. Oakland, CA: University of California Press.

Overland, Indra. 2019. "The Geopolitics of Renewable Energy: Debunking Four Emerging Myths." *Energy Research & Social Science* 49: 36–40.

Paulson, Susan. 2005. "Gendered Practices and Landscapes in the Andes: The Shape of Asymmetrical Exchanges." In *Political Ecology across Spaces, Scales, and Social Groups*, edited by Susan Paulson and Lisa L. Gezon, 174–195. New Brunswick, NJ: Rutgers University Press.

Paulson, Susan. 2017. "Degrowth: Culture, Power and Change." *Journal of Political Ecology* 24: 425–448.

Powell, Dana E. 2018. *Landscapes of Power: Politics of Energy in the Navajo Nation*. Durham, NC: Duke University Press.

Princen, Thomas, Jack P. Manno, and Pamela L. Martin, eds. 2015. *Ending the Fossil Fuel Era*. Cambridge, MA: MIT Press.

Purcell, Thomas F., and Estefania Martinez. 2018. "Post-Neoliberal Energy Modernity and the Political Economy of the Landlord State in Ecuador." *Energy Research & Social Science* 41: 12–21.

Rich, Bruce. 2013. *Mortgaging the Earth: The World Bank, Environmental Impoverishment, and the Crisis of Development*. Washington, DC: Island Press.

Santos, Boaventura de Sousa. 2014. *Epistemologies of the South: Justice against Epistemicide*. London, UK: Routledge.

Sawyer, Suzana. 2004. *Crude Chronicles: Indigenous Politics, Multinational Oil, and Neoliberalism in Ecuador*. Durham, NC: Duke University Press.

Urry, John. 2016. "Does Mobility Have a Future?" In *Mobilities: New Perspectives on Transport and Society*, edited by Margaret Grieco and John Urry, 3–20. London, UK: Routledge.

Walker, Gordon. 2021. *Energy and Rhythm: Rhythmanalysis for a Low Carbon Future*. Lanham, MD: Rowman & Littlefield.

Wang, Jundong, Zhi Tan, Jinping Peng, Qiongxuan Qiu, and Meimin Li. 2016. "The Behaviors of Microplastics in the Marine Environment." *Marine Environmental Research* 113: 7–17.

Watts, Michael J. 2012. "A Tale of Two Gulfs: Life, Death, and Dispossession along Two Oil Frontiers." *American Quarterly* 64(3): 437–467.

Weiβbach, D., G. Ruprecht, A. Huke, K. Czerski, S. Gottlieb, and A. Hussein. 2013. "Energy Intensities, EROIs (Energy Returned on Invested), and Energy Payback Times of Electricity Generating Power Plants." *Energy* 52: 210–221.

Yergin, Daniel. 2008. *The Prize: The Epic Quest for Oil, Money, and Power*. New York, NY: Free Press.

Zalik, Anna. 2015. "Resource Sterilization: Reserve Replacement, Financial Risk, and Environmental Review in Canada's Tar Sands." *Environment and Planning A: Economy and Space* 47(12): 2446–2464.

CHAPTER 8

Fields and forests

In September 2020, the Hindu nationalist majority in the Indian parliament, the Lok Sabha, passed three controversial agricultural laws. Farmers from the states of Punjab and Haryana set up camp on the outskirts of Delhi, warning that they would march on the city unless the laws were repealed. Under the call of *Dilli Chalo* (Let's go to Delhi), tens of thousands of unionised farmers blockaded the major thoroughfares and halted all transports of food and other goods into the city in January 2021. The police responded with force outside the Red Fort to secure the urban supply lines.

To understand the significance of the protests, the classic political ecology approach would be to examine them in a progressively wider geographical and denser historical context. In the decades following independence from British rule in 1947, India was plagued by famine. In the West, leading observers argued that the food shortages were an effect not of colonialism and decolonisation but the sheer number of people on the subcontinent. In *The Population Bomb*, Stanford biologist Paul Ehrlich (1968: 3) wrote that "the underdeveloped countries of the world face an inevitable population-food crisis. … While there are temporary or local reversals of this trend, it now seems inevitable that it will continue to its logical conclusion: mass starvation". India, like the planet, was running out of resources.

In the United States, agronomists and engineers developed a technical solution to the apparent overpopulation problem. The Green Revolution was rolled out internationally as part of US development aid to increase agricultural yields, but it was also framed as an alternative to the Red Revolution—a means of undermining social revolt among the world's poor (Dove and Kammen 2015: 78). In India, Punjab and Haryana were key areas for the Green Revolution's implementation. Farmers were equipped with heavy machinery and irrigation systems, and the new production system required that they purchase so-called high-yielding variety seeds and specially manufactured pesticides from large, often US-based companies. As opposed to more local economic and social relations, the Green Revolution incorporated Indian farmers in a global market of commodity exchange. To pay for agricultural inputs, they relied on credits to be repaid with income from a successful harvest. The new system, therefore, entailed an elevated degree of risk for the farmers since drought or crop failure made them economically vulnerable.

To safeguard the rural population from exploitation, the governing Congress Party regulated the agricultural markets in the 1960s. State-controlled agricultural committees acted as intermediaries between farmers and retailers, and produce had to be auctioned in markets (*mandis*) with a guaranteed minimum price. The controversial farm laws of 2020, which reflected Prime

DOI: 10.4324/9781003095521-8

Minister Narendra Modi's neoliberal economic agenda, decreed first, that produce could be sold outside the *mandis*; second, that farmers could work on contract with large corporations; and third, that the marketing of essential crops, such as onions and potatoes, would be deregulated, removing the guaranteed minimum prices. The farmers feared that the laws would leave them at the mercy of the large corporations and trap them in debt. Hence, they marched.

It comes as no surprise that the farmers blockaded the roads leading into the capital city. The United Nations (2019: 1) estimates that 55 per cent of the global population lives in cities, and in 2050, less than one-third is expected to live in rural areas. The protests, at any rate, reveal the problems associated with making a sharp distinction between the urban and the rural and point to the continuities that exist between cities and the countryside (Chapter 5). By stopping the flow of material goods into the city, the farmers impeded the urban metabolism and inhibited urban life. Indeed, if urban consumption and rural livelihoods are linked, both being products of urbanisation, it is important to question under what social, economic and environmental conditions modern life is sustained: *where* does sustainable development take place—in cities, fields or forests—and *how* is a socially just low-carbon transition spatially produced?

In addition to the rural-urban dynamic, the protests show that life on a farm cannot be explained only with reference to activities in that particular place: it must also be understood in a context of multi-scale and unequal power relations. Agricultural livelihoods are interlaced with urban consumption but also state regulation, neoliberal ideology and multinational capital. Just as the urban and the rural are intertwined, so are the local and the global (Chapter 3). In the Anglo-American tradition, political ecologists have historically been deeply concerned with issues of agriculture, forestry and land-use change. In this chapter, we explore some of the most formative debates on these topics, accounting for many key works in the traditional canon. We also look ahead to discuss how a low-carbon transition—perpetuating urban consumption patterns under more sustainable conditions—raises difficult ethical questions in the way it upsets established patterns of land use and rural livelihoods.

FROM THE TRAGEDY OF THE COMMONS TO THE TRAGEDY OF THE ENCLOSURE

It is a basic observation that most human activities take place on land. And yet, it is an observation with far-reaching political implications. Imagine a pasture like the one in Figure 8.1. More than a piece of land with particular soil and vegetation characteristics, the pasture is situated in a set of social relations that define the terms of land ownership, access and management. In a famous thought experiment, the ecologist Garrett Hardin (1968) pictured a pasture which was common property. Hardin argued that any rational person with access to the pasture would try to keep as many animals as possible on it to maximise their profit. But such "rational" behaviour, he argued, would have fateful consequences. Eventually, the pasture would be overgrazed and the ecosystem would collapse. "Therein is the tragedy", he wrote. "Each man [*sic*] is locked into a system that compels him to increase his herd without limit—in a world that is limited" (1244).

Hardin's thought experiment on overgrazing, just like Ehrlich's alarm about the inevitability of overpopulation, was taken as a warning signal by Western scholars and policymakers. Within the confines of a limited biosphere, common property regimes would lead to collapse, and this

FIGURE 8.1 Cows grazing on a mountain common in County Innlandet, Norway

Photo by Gustav Cederlöf

was not only true in the cases of overgrazed pastures and overpopulated regions but also in overfished oceans, overproduced oil fields, and overabundantly emitted CO_2 in the atmosphere. The logical solution to the "tragedy of the commons", according to Hardin and his followers, was to enclose land and other commons, turning them into private property. If privately owned, it would be in the interest of each herder to manage their pasture sustainably to ensure a long-term return. Western-dominated development agencies, such as the World Bank, also started breaking up commonly held lands in the global south to establish private property regimes. World Bank-led "development" was tantamount to enclosure, private property and commodity production (de Soto 2000). A private property regime is also intrinsic to the "New Green Revolution" recently advanced for hunger alleviation in Africa (Nyambura 2015; Moseley et al. 2018).

The communities subjected to these development programmes, however, testified about their troubling implications. On the one hand, the programmes reproduced colonially patterned power relations by which Western experts explained what marginalised communities should do in "their own" interest while generating lucrative investment opportunities for global capital (Escobar 1995). On the other, there was a concern with the premises of Hardin's tale. The tragedy of Hardin's article, Michael Goldman (1997) argued, had more to do with Hardin's own worldview—the assumptions he made and his sense of rational behaviour—than the world he attempted to describe.

Three important objections have been raised to the "tragedy of the commons" narrative. The first accepts the premises of Hardin's case but notes that there are two parts to the problem: on the one hand, there is commonly held land; on the other, profit-maximising herders who own cattle individually. Hardin maintained that the logical solution would be to break up the commons and formalise land titles. However, an equally logical case could be made for a politically contrary solution. Collective ownership of the cattle would also avert the tragedy (Harvey 2011). This would point towards a socialist rather than a capitalist property regime, and clearly, the intellectual environment in which Hardin worked at the height of the Cold War in the United States influenced the assumptions he made.

The second objection has garnered more attention and is closely associated with the political economist Elinor Ostrom's work for which she received the 2009 Nobel Prize in Economics. Like any good scholar, Ostrom (1990) took an empirical approach to the problem, and instead of starting with a thought experiment, she collected historical and contemporary evidence of how people actually managed common pool resources. Ostrom found that access to the commons, rather than reflecting a free-for-all, tended to be regulated by less formal institutions. These institutions were culturally specific (though no less "cultural" than the institution of private property) and often difficult to discern for Western experts such as Hardin, who viewed the "underdeveloped world" from his armchair. (Furthermore, while it does not necessarily impact the tragedy of the commons narrative, Hardin was known for his white nationalist views.) For anyone immersing themselves in the social context over an extended period, the rules regulating access to the commons were visible. From this perspective, a failure to manage the commons—for example from overgrazing—was an exceptional consequence of failed governance rather than an inevitable outcome. Complicating the issue further, Patricia Vondel (1987) showed how rice was grown under a Green Revolution private property regime in the swamps of South Kalimantan, Indonesia during periods of low water. But when the rice paddies flooded on a seasonal basis, they were managed as a commons for duck farming. What lands, then, were commons and what lands were privately owned? The world was far more complex than Hardin made out.

The third objection has been raised by communities, researchers and activists with lived experiences of enclosure—a process sometimes leading to entire communities' forced eviction from previously common-held lands. When the commons are turned into private property, people who depend on access to them are dispossessed of customary rights. In a damning critique, the writers' collective *The Ecologist* (1994: 106) argued that "enclosure not only destroys homes, settlements and (in many instances) livelihoods but also undermines the institutions and cultural patterns that protected 'common properties'". Taking marginalised people's lived experiences into account, the issue was less a tragedy of the commons than a tragedy of enclosure. Indeed, as the Kenyan ecofeminist Ruth Nyambura (2015) argues in relation to the New Green Revolution, the privatisation of the commons has tended to benefit large agribusiness rather than African smallholder farmers, most of whom today are rural women. Crucially, the issue has taken on renewed relevance as national and corporate actors pursue a low-carbon transition. Globally intensified efforts of forest and wildlife conservation and the development of a green economy create new frontiers of what Nancy Lee Peluso and Christian Lund (2011) define as struggles over "land control". In these struggles, powerful actors seize land and maintain access to it through symbolic and material practices of enclosure that consolidate claims to a scarce resource.

LOW-CARBON TRANSITION: LIMITS AND LAND USE CHANGE

Hardin and Ehrlich were both concerned with the consequences of human interaction with a limited resource. For Hardin, there was only one pasture, and it had fixed boundaries that the herders had to keep within. For Ehrlich, there was only a limited amount of food that an exponentially growing population could access. At some point, the resource would be depleted within that limited space or the ecosystem would collapse. Ehrlich concluded that it was necessary to impose strict population control measures in the global south to prevent overpopulation, or else mass starvation would follow. The underlying idea, as we discussed in Chapter 4, is known as neo-Malthusianism since it echoes an argument put forward by the eighteenth-century economist Thomas Malthus. In the 1980s, political ecologists derided the neo-Malthusians for apportioning blame for resource scarcity to poor people in former European colonies while lifestyles in the West were based on conspicuous consumption, drawing on resources and labour from those former colonies. Famine was not a result of a limited pool of resources but inequalities in distribution (Watts 2013).

The Danish economist Ester Boserup (1965) notably argued that rising population levels could generate economies of scale and new technology (so-called "induced innovation") so that the levels of productivity increased dramatically. In this case, depopulation was the real problem (Bryant et al. 2015). Piers Blaikie and Harold Brookfield (1987: 34) also took the neo-Malthusians to task, and drawing on a wide range of historical evidence, they concluded that "degradation can occur under rising PPR [pressure of population on resources], under declining PPR, and without PPR". However, their most important contribution was to question the use of the site—such as the pasture—as a bounded unit of analysis. As we noted in Chapter 3, Blaikie and Brookfield developed the idea of a "chain of explanation" in which human-environment interactions on a local scale had to be explained in relation to how that locality was influenced by processes acting on a regional, national and international level. When exposed to a chain of explanation, the neo-Malthusian argument fell apart.

Blaikie was particularly interested in soil erosion as a form of land degradation. Soil erosion is a naturally occurring phenomenon in which topsoil washes or blows away, yet erosion from agricultural fields is currently estimated to exceed the rate of soil formation by 10–20 per cent in the case of no tillage to more than 100 per cent under conventional tillage (IPCC 2020: 7). Heavily eroded fields are all but impossible to farm. In the neo-Malthusian narrative, soil erosion resulted from overpopulation and too high pressure on the ecosystem. But Blaikie (1985) argued that the individual site had to be exposed to a chain of explanation. A broader political economic context, defined by the social relations of production, influenced the land-use decisions made by farmers in a particular site (Figure 8.2). First, land users responded to the actions of other land users, working with available technology and knowledge. Second, they made their decisions in the context of regional and national land-tenure legislation and state policy, which structured their ability to access land and incentivised particular courses of action. Third, land users acted within the confines of an international political economy in which entire countries in the global periphery, years after political independence, remained economically dependent on single agricultural exports. It was in the interest of the national elites in these countries to produce more extensively for export, but this tended to push poor people on to degraded lands, which in a vicious cycle of marginalisation forced them to intensify production and thus even further degrade the land.

Through studies like Blaikie's, but also for example Susanna Hecht's (1985) seminal work on the political economy of deforestation, political ecology developed as a critique of analyses based on notions of absolute limits. Recently, however, climate change has reignited a debate in which the existence of absolute planetary boundaries to growth is pitted against well-established theory foiling such narratives. Advocates of degrowth (Chapter 7) often point out

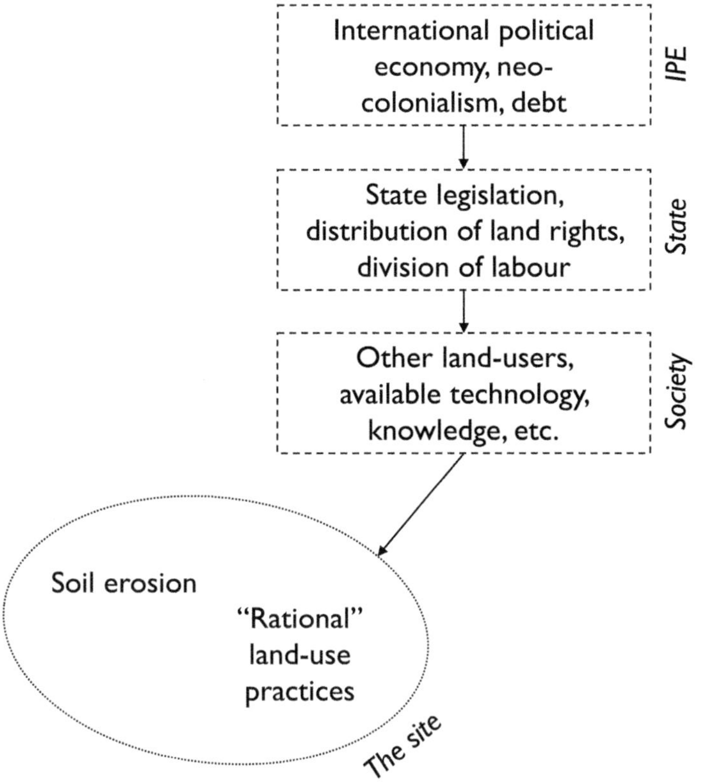

FIGURE 8.2 A chain of explanation of soil erosion, after Blaikie (1985)

that exponential economic growth, without which there is no capitalist economy, is incompatible with life on a planet with a limited resource base and a constant rate of insolation (e.g. Hickel and Kallis 2020). Any notion of "green growth" is, therefore, misguided due to the existence of absolute limits. Nevertheless, an argument like this echoes the neo-Malthusian concept of bounded space and is at odds with an analysis sensitive to the implications of scale. Clearly, the question of limits has hit a nerve.

As we write, a discussion is unfolding in the journal *Political Geography* on the theme of "Environmental limits, scarcity and degrowth". If we could pinpoint a starting point for the discussion, it began at the Second Biennial Conference of the Political Ecology Network (POLLEN) in Oslo in 2018. In a provocative keynote lecture, Paul Robbins argued that political ecologists repeatedly have shown how "scarcity", "limits" and "austerity" are ideological concepts that tend to be imposed on people by an elite whose interests they serve. As a result, narratives of natural limits may work contrary to an emancipatory politics and a socially just low-carbon transition. In response, Erik Gómez-Baggethun asserted that it is dangerous in a time of rapid climate change to relativise well-established research on the existence of planetary boundaries. Furthermore, if self-imposed, notions of austerity, scarcity and self-limitation can be the starting points for a politics in which an environmentally conscious subject does things differently—a way of life in which societies prioritise things other than economic growth. Robbins' and Gómez-Baggethun's contributions, which were later published (Gómez-Baggethun 2020; Robbins 2020), now set the frames for an exciting debate.

Two things are important to take account of here. First, while the neo-Malthusians spoke of absolute limits to growth, the question in the context of climate change is about the risk of overstepping critical systemic "thresholds". While always in a state of flux, ecosystems and the climate system tend to develop into relatively steady states that persist even as they are exposed to disturbance. The global climate in the Holocene period can be seen as one such relatively steady state. However, at some point, a threshold can be passed due to systemic stress, which leads to abrupt, non-linear environmental change (Zimmerer 2000). As Carl Cassegård (2021: 4, original emphasis) argues, "Surely, it can't possibly make any difference to the global climate whether we release an additional CO_2 molecule into the air or not, or whether we take the car to work or not. It's precisely because we are correct [in thinking so] that we will be so *surprised* when the catastrophe arrives." Indeed, in a provocative statement drawing on the philosopher Hegel, he goes on to note that "when the dialectical *Umschlag* finally occurs, we realize that the catastrophe has been going on all the time, and that we were in its midst even when things still seemed fine." The risk in the case of climate change is less that the planet runs out of resources than that critical thresholds are transgressed, making human life on the planet difficult.

Second, the costs and benefits of gradual or abrupt environmental change tend to be unequally distributed among social groups. Climate change transforms the environmental circumstances for human life so that the livelihoods of the already resource-weak, who are often forced to extract a living from marginal lands, become even more difficult to maintain. It also increases the vulnerability of marginalised communities to hazards, such as wildfires and floods, putting them at an elevated degree of risk (Wisner et al. 2004). Just as the impacts of global environmental change are unevenly spread, the costs and benefits of efforts to mitigate and adapt to climate change are distributed unequally along social gradients of class, race, gender and caste (Paprocki 2021). Notably, the land-use changes provoked by a low-carbon transition constitute a key area in which these inequalities are negotiated and reinforced.

In its Special Report on Climate Change and Land, the IPCC (2020) concludes that to reduce the global carbon emission levels to the extent necessary for staying below the 1.5°C target set in the Paris Agreement, the patterns of human activities on land must change. In the report's technical language, "All assessed modelled pathways that limit warming to 1.5°C or well below 2°C require land-based mitigation and land-use change, with most including different combinations of reforestation, afforestation, reduced deforestation, and bioenergy" (24). To unpack this statement, it is crucial to understand energy use and carbon capture as spatially constituted phenomena—phenomena that give rise to particular geographical patterns on land. Take fossil fuels, for example. Fossil fuels form from biomass that has grown on land, or from algae that have covered the sea, and that over millennia have decomposed under high pressures and temperatures. When we burn a fossil fuel—say, driving a car—we are in other words burning an ancient accumulation of biomass that has grown on land. To use the environmental historian Rolf Peter Sieferle's (2001) memorable phrase, fossil fuels are a "subterranean forest" that we tap into from above.

By contrast, renewable energy sources take up contemporary land. If we replace petrol with a biofuel to run a car, we indirectly also need to replace the land historically accumulated in petrol with crops and technologies that exist in fields and forests today. There is a vast difference between the amount of land needed to produce one measure of fossil energy compared to one measure of renewable energy. The energy scholar Vaclav Smil (2015) has developed the concept of "power density" to explore the spatial extensiveness of different energy sources. Power density is defined as "watts per square metre" (W/m^2), and Smil estimates that if you cover one square metre of land with photovoltaic solar panels, you can on average harness 10–20 W. At a coal seam, in comparison, such as that tapped at China's Heidaigou Mine, he reports a power density falling just short of 1,000 W/m^2. Thus, you can extract 50–100 times more energy at a coal seam than from an equivalent area covered with solar panels. In the North Antelope Rochelle Mine in Wyoming, USA—one of the world's largest open-pit coal mines—the power density reaches a whopping 12,000 W/m^2. There is a dizzying difference, then, in the land intensity of renewable compared to fossil energy production.

Building on Smil's concept, political ecologists are identifying the geographical reconfigurations underpinning a low-carbon transition as some of the most pressing issues today: if we transition from a subterranean energy regime based on extraction from oil wells and coal mines to a land-based energy regime, where energy is harnessed horizontally from biofuel plantations and solar parks, *whose* land should we use for energy production? And to sustain *whose* consumption? Spurred by the urban demand for energy, a low-carbon transition thereby transforms the circumstances for rural livelihoods on a global scale, often through new enclosures and dispossessions in a process of "green grabbing" (Fairhead et al. 2012; cf. Huber and McCarthy 2017; Hornborg et al. 2019). Yet the land-use changes involved not only have political implications but also give rise to difficult ethical dilemmas, evoking concepts of environmental justice and injustice. On what grounds is it morally right that some communities—often indigenous, low-caste or poor—should make space for the energy production of others?

During the UN Conference on Climate Change (COP26) in Glasgow in 2021, India undertook to source 50 per cent of its energy needs from renewables by 2030. As part of the country's earlier National Mission on Biodiesel (NMBD), the evergreen shrub *Jatropha curas* was singled out as a particularly promising biofuel for the green economy, and the Government of India sponsored large Jatropha plantations in areas classified by the state as "wastelands". In an important study, Jennifer Baka (2017) demonstrates how powerful narratives of "wastelands"—argued to be empty, idle spaces that can be "improved" to help mitigate climate change—widely legitimised the mission. However, on closer inspection, the targeted lands were far from empty. Baka finds that they were used by local landless communities who made charcoal from the ubiquitous tree *Proposis juliflora* and often extracted more useful energy from Proposis than would be possible from Jatropha. Through the NMBD, common lands were turned into private property for the purpose of "green" commodity production, and Baka shows how the enclosures erased existing energy economies and livelihoods in rural communities. The green transition led to "energy dispossessions" through the land-use changes it triggered. The energy transition can be seen as one instance of how climate change adaptation can redistribute power and resources away from already vulnerable communities through agrarian change (Paprocki 2021).

GOVERNING THE FOREST

Nalgaon is a village on the slopes of the Indian Himalayas. It is easy to appreciate the importance of the forest to village life: Nalgaon's houses are built from stone and clay (but increasingly also from rebar and cement); the doors are wooden and the roofs are made of slate. Mainly women live in the village, because the men have long migrated to cities or enlisted in the army on the Indo-Chinese border in search of a family income. The women spend many hours a day in the forest grazing their cattle and collecting firewood. The cattle give milk, and their manure fertilises the village's terraced soils. The forest shapes human life while the villagers and their animals shape the forest ecosystem.

The forests around Nalgaon are dominated by a pine species known as chir (*Pinus roxburghii*; Figure 8.3). For an attentive observer, it is a remarkable environment. The local communities find almost no use in pine—at best, the needles can be mixed with manure and be used for

FIGURE 8.3 Chir pine forest in the Central Himalayas

In Linnean taxonomy, chir pine is known as *Pinus roxburghii*. It was named after the Scottish botanist William Roxburgh who worked in British India. Botanists like Roxburgh played an important role in the colonial project. By systematising the natural environment, they produced a particular kind of knowledge (Chapter 2), which made it possible to manage forests scientifically and maximise yields in the timber industry. This was a kind of knowledge that marginalised non-Western epistemologies of the forest.

Photo derivative of Rick Bradley, CC-BY-2.0

mulching. As we saw in Chapter 2, the British enclosed India's forests in the nineteenth century, and most of the enclosures became "reserved forest", property of the state's Forest Department. For the colonial administration and later postcolonial governments, the forest was a resource for economic development to be managed sustainably towards that end. In keeping with the market logic guiding the timber industry, the value of the forest was only realised when trees were cut and timber sold. Growing tall and straight, chir provided excellent timber. The hallmarks of the enclosures were the Indian Forest Acts of 1878 and 1927 when, as Madhav Gadgil and Ramachandra Guha (1992: 134) write, "by one stroke of the executive pen, [the British] attempted to obliterate centuries of customary use of the forest by rural populations all over India".

For the village communities, the enclosures meant that by carrying on doing what they normally did, they broke the law: they had become trespassers on state land. Almost a century later, the Indian Forest Act of 1927 continues to regulate India's forests, and in Nalgaon, the villagers can only access forest pastures and materials through formal or informal agreements with local forest guards. Beyond questions of access, however, the ecological transformations occurring in the wake of the enclosures have also marginalised the villagers in a more camouflaged sense. In 1793, a visitor to the area noted that "the forests of oak, fir, and boorans [rhododendron] are here more extensive and the trees of greater magnitude, than any I have ever seen" (T. Hardwicke citied in Guha 1989: 35). Yet the Forest Department's sustained planting of chir meant that, over time, the once biodiverse broad-leaf forest transmuted into a pinewood plantation. Noted for its banj oak (*Quercus leucotrichophora*), the broad-leaf forest served local needs, and it was managed to provide fodder, pastures and materials.

In their path-breaking work in environmental history, Gadgil and Guha (1992) depicted the broad-leaf forest in rosy terms: the precolonial period saw an ecosystem in balance as well as nonconflictual use of common pool resources. However, feminist scholars have provided an important corrective to such accounts. Bina Agarwal (1998) notably exposed the risks associated with romanticising particular periods of the past and showed how gender, class and caste divisions regulated access to the forest commons in socially unequal ways in the time before the enclosures. Thus, while colonialism disrupted customary practices and institutions—integrating fields and forests in a globalised extractive economy—we must also maintain a critical stance towards precolonial and indigenous access arrangements to natural resources.

Like all cases, the Himalayan case is historically and geographically unique. But it illustrates a key point that goes beyond the specificity of the individual example. Political ecologists infer that forest ecosystems embody economic, social and cultural values, and the terms under which a forest is governed are, therefore, inherently political. In *The Social Lives of Forests*, Hecht et al. (2014: 5) write:

> What lies under, behind, or in the past of … forests are their long-term legacies of human action: they are artifacts of human ideas about nature, expressions of economies, and places mediated by institutions and their materialized practices[,] as much as they are habitats evolving in response to evolution and planetary processes.

As a result, if we are to appreciate the complexity of issues such as deforestation, forest degradation and forest conservation—for example in the context of climate change adaptation—it is crucial to approach forests, on the one hand, as socio-natural spaces, and on the other, as historical environments in which unequal and multi-scalar power relations are at work.

FROM CONFLICT TO ENVIRONMENTALITY

One way to understand the politics that take place among trees is to study forests in terms of conflict and social struggle. In *The Unquiet Woods*, a landmark study for South Asian political ecology, Guha (1989) drew attention to how the enclosure and ecological transformation of the Himalayan forests occurred under periods of great social unrest. Guha wrote in a tradition of "history from below", exploring the agency of marginalised people, and his starting point was an environmental movement. While many sociologists and political scientists long assumed that conservation and environmental issues were the preserves of affluent people (e.g. Inglehart 1995), this shed light on the "environmentalism of the poor"—environmental struggles engaged in by vulnerable communities whose livelihoods are undermined by environmental degradation (Guha and Martínez-Alier 1997).

The Chipko Andolan gained widespread international attention in the 1970s through accounts of its "tree hugging" women members (Mawdsley 1998). Many interpreted Chipko in ecofeminist terms (Chapter 4), suggesting that due to their mutual exploitation in society, a natural bond existed between women and the forest. Supposedly, the act of tree-hugging expressed this bond and feminine "care". Through detailed empirical analysis, Guha rebutted such accounts, which essentialised both women and the forest. Instead, he demonstrated how Chipko protests were rooted in the long-term degradation of the forest ecosystem; a kind of degradation that undermined village livelihoods while serving the interests of commercial forestry. Chipko, Guha argued, was but the latest instantiation of a much longer history of struggle from below against the colonial and postcolonial state.

One particularly violent episode attests to this prolonged struggle. When the British Raj enclosed the forests, the villagers responded violently. In the 1910s and 1920s, they repeatedly set the forested landscape ablaze. In 1921, an unusually dry year, a vast 1,100 km^2 burnt in Kumaon Province alone (Guha 1989: 52). But the fires did not rage indiscriminately. They occurred almost exclusively in the chir pine plantations worked for timber and resin, while broad-leaf forests were spared. In his study, Guha assesses the scale of ecological change, but he also combines this environmental focus with an actor-based analysis of subaltern protest. For him, the forest fires were a tactical means for a marginalised actor to assert power over a dominant actor—just as tree-hugging (or in the present, farmers' roadblocks) is a nonviolent tactic. However, in a more recent study, Arun Agrawal (2005) takes a different approach, which has stoked intense debate among political ecologists. Agrawal's *Environmentality* attends to the same history as Guha's *The Unquiet Woods*, but it does so based on a different, Foucauldian understanding of power (Chapter 2). If we read the two studies side-by-side, we can gain a deeper understanding of the problems associated with environmental governance, but also how the kinds of research questions asked by political ecologists have become more refined over time.

Agrawal notes how the Kumaoni villagers set fire to the forest in the 1910s, but he also points out that by the 1990s, they had become assiduous forest conservationists. As one of his interlocutors states, "If we don't safeguard the forest, who else will?" (2). Agrawal's objective is to explain the mechanisms underlying this remarkable change in attitude. To understand his argument, we need to make a conceptual detour and, in doing so, go beyond an analysis in which actors exist as fixed, already-given subjects: "The relationships of subjects to the environment ... need to be examined in their emergence", as Agrawal writes, "not simply [be] taken as part of a larger politics by preexisting interests" (211). Developing a Foucauldian perspective on power, Agrawal argues that people's understanding of who they are—their subjectivity—changes at the same time as they modify their interactions with the environment within a particular institutional setting. An actor, then, does not simply exist but emerges historically.

Towards the end of Chapter 2, we saw how power and knowledge are intimately tied to one another. One of the great insights of late twentieth-century philosophy was that the words we use to describe the world do not map neatly on to it, but rather, material objects, bodies, human practices and "nature" itself come into being as meaningful entities through the knowledge we produce about them. Knowledge, in turn, is always produced in a historical context in which it reflects situated interests, priorities and purposes. "Power" in this way flows through knowledge and the ways in which we conceive of the world.

Foucault developed a range of concepts to tease out the implications of such a notion of power. For political ecologists, "governmentality" has become a particularly fruitful concept (see notably Wang 2015; Fletcher and Cortes-Vazquez 2020). We earlier witnessed how a certain kind of knowledge emerged in the eighteenth century as part of European colonialism, classifying human bodies in different races. Knowledge about European racial superiority thereby allowed the colonisers, and to some extent also the colonised, to make sense of the world they encountered as well as their own place in it (Said 1979). Crucially, this racialised knowledge was reproduced in colonial institutions of government, trade, science, art and literature. Foucault also argues that modern government only becomes possible through the production of knowledge about a population: knowledge about its characteristics, behaviour and development brings the population into being as a meaningful entity, making it governable. This is also what Stephanie Rutherford (2007: 293) concludes in a review on "green governmentality": "Demographics, statistics and insurance established new power/knowledge formations that placed the individual into a larger population, whose characteristics could be measured, assessed and managed. Making both the individual and population objects of knowledge for government is the hallmark of modern rule".

Taking one further step, Foucault contends that power, rather than being centrally located with a sovereign actor, is distributed in a multiplicity of sites—in organised human practices. The concept of governmentality intends to capture how people come to understand themselves and modify their behaviour in accordance with the power/knowledge formation they live in through various social mechanisms. People internalise government aims that define what it means to be a good citizen. For instance, when racialised people were exposed to colonial power/knowledge, they were also called upon to engage in their own management as racialised subjects. Ultimately, Foucault argues that people come to self-regulate, seeking to comply with the rationality of government, since they understand this rationality to be normal and right. Government seeks to shape the "conduct of conduct", he writes, producing subjects able to comply with its rationality.

In his study on forest governance, Agrawal coins the term "environmentality" to denote "why, when, how, and in what measure people come to develop an environmentally oriented subject position" (3). He then shows how the colonial state reacted to the forest fires in Kumaon by decentralising the institutions of forest management. By establishing village-based forest councils (so-called *van panchayats*), decision-making power was dispersed in a multiplicity of sites, enrolling village communities in institutions of environmental governance. The forest councils operated within a general set of guidelines, and these called upon the council members to monitor ecological changes via indicators developed by the Forest Department. Gradually, the villagers internalised certain conservationist aims as their own, and they started identifying as an environmentally conscious subject. In an important critique of Agrawal's work, Neera Singh (2013: 190) argues that "contrary to the expectation that actions follow beliefs, actions—such as participating in the governing of the environment—often lead to new beliefs and, thus, new subjectivities". In the Indian Himalayas, governance of the enclosed forests rested on the production of a new subjectivity, which identified with the aims of dominant political economic interests.

LOW-CARBON TRANSITION: CARBON CAPTURE AND SUBJECT FORMATION

Broadly speaking, forests are important for a range of reasons: they are habitats for wildlife and spaces of great biodiversity; they also provide timber for economic development. They are sources of subsistence for local communities, and they are spaces of cultural attachment, identity and spirituality. Not least, they are the planet's most important carbon sinks. Absorbing CO_2 from the atmosphere, forests have been essential for maintaining a stable global climate during the Holocene. Deforestation and forest degradation are human-induced processes that contribute to a massive release of CO_2 back into the atmosphere (Box 8.1). The IPCC (2020: 12) estimates that about 13 per cent of all anthropogenic CO_2 emissions result from deforestation itself, which releases carbon embedded in biomasses and soils.

Since the early 1990s, forests—especially in the global south—have become important spaces in political attempts to reduce the atmospheric carbon levels. In the wake of the 2015 Paris Agreement, each country now sets a target for itself, known as Nationally Determined Contributions (or NDCs), to that end. Although the target is non-binding, the most immediate way for a country to achieve its NDCs is to reduce the levels of emissions within its own borders. However, the Paris Agreement also permits countries to continue emitting CO_2 above their set targets if they compensate for it with investments aiming to reduce emissions in another country. The logic behind this is the same as that which governs the Payment for Ecosystem Services paradigm that we explored in Chapter 4. In practice, in the context of CO_2 trading, this means that the environmental burden of consumption in the global north can be displaced (or "offset") to the global south while potentially providing development and conservation benefits there (Chapter 9).

The Paris Agreement establishes a market in which CO_2 is given a price tag so that emissions in one place can be neutralised by investments elsewhere. The Sustainable Development Mechanism, as the market is called, emulates the Clean Development Mechanism (CDM) which has existed since 1997 under the Kyoto Protocol. Through the CDM, countries seeking to curb their emissions can purchase a "carbon credit". A carbon credit represents an investment

Box 8.1 Glossary of common forestry-related terms

Afforestation	the conversion of land into forest, often for commercial exploitation or carbon capture, which can be linked to the commodification of a particular aspect of the forest, such as its role as a carbon sink
Conservation	the protection of forests and wildlife, for example through the construction of national parks. Conservation has often occurred at the expense of poor forest-dependent communities who are evicted and denied access to protected areas
Deforestation	the clearing of a large, forested area for economic activities such as agriculture, cattle ranching, mining or timber use, often to meet the demands of urban consumption
Degradation	the reduction in the density, biodiversity and structure of a forest, which tends to undermine the livelihoods of poor forest-dependent communities and nonhuman species
Fragmentation	the breakup of a swathe of forested land into multiple islands, for example through road construction, agriculture or mining. Fragmentation reduces the size of animal habitats and makes pastoralist livelihoods more difficult to maintain
Reforestation	the restocking of a previously deforested or degraded forest
Resurgence	the large-scale recovery and regeneration of forests

in a project in the global south that, for instance, aims to increase the energy efficiency of an economic activity or improve soil management and thus reduce CO_2 emissions. Parallel to the CDM, a voluntary market for carbon offsets has been established so that individuals and firms can opt to compensate for the burden of their consumption. Many have come across this option when purchasing a flight ticket, usually framed in a powerful discourse asserting the "carbon neutrality" of their actions (Benjaminsen and Svarstad 2021).

One way for governments to offset their NDCs is to invest directly (through a grant) or indirectly (through the carbon credit market) in a programme run by the United Nations known as "Reducing Emissions from Deforestation and Forest Degradation" or, for short, REDD+. The programme transfers capital to forest conservation projects in the global south in which communities undertake *not* to cut down their forests. In this way, CO_2 emissions from consumption are neutralised by emissions that, in the absence of REDD+, would have occurred from deforestation. Importantly, the programme does not counteract historical CO_2 emissions but only offsets the impact of future economic growth. The "+" in REDD+ was added by the signatories to represent the additional development and conservation benefits that could result from the programme.

Political ecologists have studied REDD+ closely and come to four conclusions that attest to the forest as a space of conflict as well as environmentality. First, the programme is focused on "forest conservation", but no consensus exists on the definition of a forest. "[A]lthough landscapes and forests obviously exist biotically", as Hecht (2014: 99) argues, "what exactly constitutes a 'true' or 'authentic' forest or what even is counted as forest is controversial". And yet, REDD+ enforces particular definitions which serve its purposes and gloss over these ontological conflicts. As one study from Indonesia shows, the recorded rate of deforestation differed by up to 27 per cent in the study area depending on whether Indonesia's national forest definition, the Food and Agricultural Organisation's definition, or a third definition often used in REDD+ projects was used (Romijn et al. 2013). How the conceptual boundaries around a forest (or other object) are drawn, therefore, has far-reaching epistemological implications, and the choice of a particular "system boundary" reflects a cultural worldview and contingent political interests (Cederlöf and Hornborg 2021).

Second, just as the colonial enclosures in the Himalayas in the nineteenth century, REDD+ can lead to new enclosures of common lands. As the purpose of the programme is to restrict deforestation, forest conservation areas must be clearly delineated. Ownership must also be plainly defined and access arrangements be formally regulated. When forests in Asia, Africa and Latin America again are enclosed for the benefit of geographically remote consumption, it is not far-fetched to ask, as Adam Bumpus and Diana Liverman (2011) do, if the CDM and REDD+, in fact, constitute a form of "carbon colonialism" through green grabbing.

Third, REDD+ enrols forests in a process of commodification. As a commodity that can be traded, the forest is assigned economic value based on how much CO_2 it has the potential to capture (or fail to emit, to be precise). Thus, the programme makes spaces available for investments that curb CO_2 emissions at the lowest possible cost. These investments counteract the need to alter the economic drivers and investment patterns that generate emissions in the first place. However, the commodification of forests based on their importance as carbon sinks devalues other biophysical aspects of the forest, just as it obscures the significance of forest access to local livelihoods (Leach and Scoones 2015).

Finally, recent studies that take a Foucauldian approach to REDD+ demonstrate that for the programme to work, it must successfully produce subjects willing to take part in forest conservation. A project in Nigeria's Cross River engaged the Ekuri forest community, and Adeniyi Asiyanbi et al. (2019) argue that it did so through different forms of environmentality. Ekuri community has historically managed its forests through indigenous institutions, and on several occasions, its members have mobilised collectively to challenge different forms of domination—as Asiyanbi et al. argue, "to maintain a somewhat collective, relational autonomy" (133). The Ekuri nevertheless came to internalise REDD+ aims in three main ways.

First, REDD+ promises its target communities financial compensation for taking part in forest conservation. In Cross River, this spoke to Ekuri community's aspiration for a better life, and Asiyanbi et al. show how REDD+ appeared as an object of desire for a "subject of hope". Next, the programme operates through a disciplinary form of environmentality in which a moral burden, often unfairly, is placed on communities such as Ekuri: if they do not participate, they are vilified as a community driving deforestation. REDD+ thereby produced a "moral subject", which if it complied with the programme aims, would identify as a "good" forest steward. Lastly, by adopting an identity as a conservationist community, Ekuri were able to challenge other forms of domination. For example, when a large road construction project was planned in the region—causing deforestation but also threatening Ekuri autonomy—the community mobilised as a conservationist subject backed by the internationally sanctioned aims of REDD+. The programme thus allowed them to act as an "unruly subject". Multiple forms of environmentality, then, operated in complex and often overlapping ways in Cross River, and through their study, Asiyanbi et al. contribute to a larger debate in which the definitions of "multiple environmentalities" are being established (Fletcher 2017; Fletcher and Cortes-Vazquez 2020).

CONCLUSION AND LEARNING OUTCOMES

It is no understatement that fields and forests are fiercely contested socio-natural spaces. The decisions of farmers, foresters and forest-dependent communities shape these spaces, but their decisions cannot be understood without taking long-term historical trajectories and multi-scale processes into account. In this chapter, we have explored some of the most formative debates in Anglo-American political ecology related to agricultural and forest-linked livelihoods. Perhaps the most central question is how land tenure and access should be regulated; what the merits of common, private, state and unregulated property rights are—and for whom. Furthermore, there is currently disagreement in the field on whether absolute planetary boundaries set limits to growth when forceful critiques of neo-Malthusianism have lain at the heart of it since its inception.

Regardless of the position one takes on property rights and limits, a low-carbon transition raises challenging ethical questions. The spatial extensiveness of renewables and forests assigned for carbon capture—in contrast to the immense power densities of fossil fuels—means that a green economy will upset the land-use patterns developed by fossil capital. This creates new enclosures and dispossessions with socially unjust outcomes. A low-carbon transition, then, puts the spotlight on the governance of fields and forests, and how different modes of power operate in these landscapes.

Having read this chapter, we hope that you can:

- Interpret longstanding debates on commons and enclosures, as well as the economic interests and lived experiences they express
- Recognise neo-Malthusian narratives and apply a "chain of explanation" to refute them
- Explain why forests must be seen as socio-natural spaces shaped by long-term histories of social struggle and conflictual economic priorities
- Account for the concept of "environmentality" and how subjects form in the process of human-environment interaction within a given institutional setting

FURTHER READING

Baka, Jennifer. 2017. "Making Space for Energy: Wasteland Development, Enclosures, and Energy Dispossessions." *Antipode: A Radical Journal of Geography* 49(4): 977–996.

This paper is a brilliant account of the contested politics of land use and enclosure associated with biofuel production and agrarian change. It provides a rich conceptual discussion that adds to this chapter, as well as a detailed case study of renewable energy production from Tamil Nadu, India.

Hecht, Susanna B., Kathleen D. Morrison, and Christine Padoch, eds. 2014. *The Social Lives of Forests: Past, Present, and Future of Woodland Resurgence.* Chicago, IL: University of Chicago Press.

This edited collection is a smorgasbord of concepts and case studies from many of the most notable political ecologists working on forestry-related issues. It offers excellent further reading—in parts or its entirety—on issues of deforestation, forest degradation and fragmentation, but it is also noteworthy for the way it challenges many common assumptions and provides a positive narrative of forest resurgence.

BIBLIOGRAPHY

Agarwal, Bina. 1998. "Environmental Management, Equity and Ecofeminism: Debating India's Experience." *Journal of Peasant Studies* 25(4): 55–95.

Agrawal, Arun. 2005. *Environmentality: Technologies of Government and the Making of Subjects.* Durham, NC: Duke University Press.

Asiyanbi, Adeniyi P., Edwin Ogar, and Oluyemi A. Akintoye. 2019. "Complexities and Surprises in Local Resistance to Neoliberal Conservation: Multiple Environmentalities, Technologies of the Self and the Poststructural Geography of Local Engagement with REDD+." *Political Geography* 69: 128–138.

Baka, Jennifer. 2017. "Making Space for Energy: Wasteland Development, Enclosures, and Energy Dispossessions." *Antipode: A Radical Journal of Geography* 49(4): 977–996.

Benjaminsen, Tor A., and Hanne Svarstad. 2021. *Political Ecology: A Critical Engagement with Global Environmental Issues.* Cham, Switzerland: Palgrave Macmillan.

Blaikie, Piers. 1985. *The Political Economy of Soil Erosion in Developing Countries.* London, UK: Longman.

Blaikie, Piers, and Harold Brookfield, eds. 1987. *Land Degradation and Society.* London, UK: Methuen.

Boserup, Ester. 1965. *Conditions of Agricultural Growth: The Economics of Agrarian Change under Population Pressure.* London, UK: Earthscan.

Bryant, Raymond L., Ángel Paniagua, and Thanasis Kizos. 2015. "Governing People in Depopulated Areas." In *International Handbook of Political Ecology*, edited by Raymond L. Bryant, 233–245. Cheltenham, UK: Edward Elgar.

Bumpus, A. G., and D. M. Liverman. 2011. "Carbon Colonialism? Offsets, Greenhouse Gas Reductions, and Sustainable Development." In *Global Political Ecology*, edited by Richard Peet, Paul Robbins, and Michael J. Watts, 203–224. London, UK: Routledge.

Cassegård, Carl. 2021. *Toward a Critical Theory of Nature: Capital, Ecology, and Dialectics.* London, UK: Bloomsbury Academic.

Cederlöf, Gustav, and Alf Hornborg. 2021. "System Boundaries as Epistemological and Ethnographic Problems: Assessing Energy Technology and Socio-Environmental Impact." *Journal of Political Ecology* 28: 111–123.

de Soto, Hernando. 2000. *The Mystery of Capital: Why Capitalism Triumphs in the West and Fails Everywhere Else.* London, UK: Black Swan.

Dove, Michael R., and Daniel M. Kammen. 2015. *Science, Society and the Environment: Applying Anthropology and Physics to Sustainability.* London, UK: Routledge.

Ehrlich, Paul R. 1968. *The Population Bomb.* New York, NY: Ballantine Books.

Escobar, Arturo. 1995. *Encountering Development: The Making and Unmaking of the Third World.* Princeton, NJ: Princeton University Press.

Fairhead, James, Melissa Leach, and Ian Scoones. 2012. "Green Grabbing: A New Appropriation of Nature?" *Journal of Peasant Studies* 39(2): 237–261.

Fletcher, Robert. 2017. "Environmentality Unbound: Multiple Governmentalities in Environmental Politics." *Geoforum* 85: 311–315.

Fletcher, Robert, and Jose A. Cortes-Vazquez. 2020. "Beyond the Green Panopticon: New Directions in Research Exploring Environmental Governmentality." *Environment and Planning E: Nature and Space* 3(2): 289–299.

Gadgil, Madhav, and Ramachandra Guha. 1992. *This Fissured Land: An Ecological History of India*. New Delhi, India: Oxford University Press.

Goldman, Michael. 1997. "'Customs in Common': The Epistemic World of the Commons Scholars." *Theory and Society* 26(1): 1–37.

Gómez-Baggethun, Erik. 2020. "More Is More: Scaling Political Ecology within Limits to Growth." *Political Geography* 76: 102095.

Guha, Ramachandra. 1989. *The Unquiet Woods: Ecological Change and Peasant Resistance in the Himalaya*. New Delhi, India: Oxford University Press.

Guha, Ramachandra, and Juan Martínez-Alier. 1997. *Varieties of Environmentalism: Essays North and South*. London, UK: Earthscan.

Harvey, David. 2011. "The Future of the Commons." *Radical History Review* 109: 101–107.

Hecht, Susanna B. 1985. "Environment, Development and Politics: Capital Accumulation and the Livestock Sector in Eastern Amazonia." *World Development* 13(6): 663–684.

Hecht, Susanna B. 2014. "The Social Lives of Forest Transitions and Successions: Theories of Forest Resurgence." In *The Social Lives of Forests: Past, Present, and Future of Woodland Resurgence*, edited by Susanna B. Hecht, Kathleen D. Morrison, and Christine Padoch, 97–113. Chicago, IL: University of Chicago Press.

Hecht, Susanna B., Kathleen D. Morrison, and Christine Padoch. 2014. "From Fragmentation to Forest Resurgence: Paradigms, Representations, and Practices." In *The Social Lives of Forests: Past, Present, and Future of Woodland Resurgence*, edited by Susanna B. Hecht, Kathleen D. Morrison, and Christine Padoch, 1–8. Chicago, IL: University of Chicago Press.

Hickel, Jason, and Giorgos Kallis. 2020. "Is Green Growth Possible?" *New Political Economy* 25(4): 469–486.

Hornborg, Alf, Gustav Cederlöf, and Andreas Roos. 2019. "Has Cuba Exposed the Myth of 'Free' Solar Power? Energy, Space, and Justice." *Environment and Planning E: Nature and Space* 2(4): 989–1008.

Huber, Matthew T., and James McCarthy. 2017. "Beyond the Subterranean Energy Regime? Fuel, Land Use and the Production of Space." *Transactions of the Institute of British Geographers* 42(4): 655–668.

Inglehart, Ronald. 1995. "Public Support for Environmental Protection: Objective Problems and Subjective Values in 43 Societies." *PS: Political Science and Politics* 28(1): 57–72.

IPCC. 2020. *Climate Change and Land: An IPCC Special Report on Climate Change, Desertification, Land Degradation, Sustainable Land Management, Food Security, and Greenhouse Gas Fluxes in Terrestrial Ecosystems*. Cambridge, UK: Cambridge University Press.

Leach, Melissa, and Ian Scoones, eds. 2015. *Carbon Conflicts and Forest Landscapes in Africa*. Abingdon, UK: Earthscan.

Mawdsley, Emma. 1998. "After Chipko: From Environment to Region in Uttaranchal." *Journal of Peasant Studies* 25(4): 36–54.

Moseley, William G., Matthew A. Schnurr, and Rachel Bezner-Kerr, eds. 2018. *Africa's Green Revolution: Critical Perspectives on New Agricultural Technologies and Systems*. London, UK: Routledge.

Nyambura, Ruth. 2015. "Agrarian Transformation(s) in Africa: What's in It for Women in Rural Africa?" *Development* 58(2–3): 306–313.

Ostrom, Elinor. 1990. *Governing the Commons: The Evolution of Institutions for Collective Action*. Cambridge, UK: Cambridge University Press.

Paprocki, Kasia. 2021. *Threatening Dystopias: The Global Politics of Climate Change Adaptation in Bangladesh*. Ithaca, NY: Cornell University Press.

Peluso, Nancy Lee, and Christian Lund. 2011. "New Frontiers of Land Control: Introduction." *Journal of Peasant Studies* 38(4): 667–681.

Robbins, Paul. 2020. "Is Less More… or Is More Less? Scaling the Political Ecologies of the Future." *Political Geography* 76: 102018.

Romijn, Erika, John Herbert Ainembabazi, Arief Wijaya, Martin Herold, Arild Angelsen, Louis Verchot, and Daniel Murdiyarso. 2013. "Exploring Different Forest Definitions and Their Impact on Developing REDD+ Reference Emission Levels: A Case Study for Indonesia." *Environmental Science & Policy* 33: 246–259.

Rutherford, Stephanie. 2007. "Green Governmentality: Insights and Opportunities in the Study of Nature's Rule." *Progress in Human Geography* 31(3): 291–307.

Said, Edward W. 1979. *Orientalism: Western Conceptions of the Orient*. New York, NY: Vintage Books.

Sieferle, Rolf Peter. 2001. *The Subterranean Forest: Energy Systems and the Industrial Revolution*. Cambridge, UK: White Horse Press.

Singh, Neera M. 2013. "The Affective Labor of Growing Forests and the Becoming of Environmental Subjects: Rethinking Environmentality in Odisha, India." *Geoforum* 47: 189–198.

Smil, Vaclav. 2015. *Power Density: A Key to Understanding Energy Sources and Uses*. Cambridge, MA: MIT Press.

The Ecologist. 1994. "Whose Common Future: Reclaiming the Commons." *Environment & Urbanization* 6(1): 106–130.

United Nations. 2019. *World Urbanization Prospects 2018: Highlights*. New York, NY: United Nations.

Vondel, Patricia. 1987. "The Common Swamplands of Southeastern Borneo: Multiple Use, Management, and Conflict." In *The Question of the Commons: The Culture and Ecology of Communal Resources*, edited by Bonnie J. McCay and James M. Acheson, 231–249. Tucson, AZ: University of Arizona Press.

Wang, Ting-jieh. 2015. "Green Governmentality." In *International Handbook of Political Ecology*, edited by Raymond L. Bryant, 318–331. Cheltenham, UK: Edward Elgar.

Watts, Michael J. 2013 (1983). *Silent Violence: Food, Famine, and Peasantry in Northern Nigeria*. Athens, GA: University of Georgia Press.

Wisner, Ben, Piers Blaikie, Terry Canon, and Ian Davis. 2004. *At Risk: Natural Hazards, People's Vulnerability and Disasters*, 2nd ed. London, UK: Routledge.

Zimmerer, Karl S. 2000. "The Reworking of Conservation Geographies: Nonequilibrium Landscapes and Nature-Society Hybrids." *Annals of the Association of American Geographers* 90(2): 356–369.

CHAPTER 9

Virtual political ecology

Tate Modern is one of London's premier art galleries. Located in an old power station, the museum commissions a large installation each year to be displayed in the former turbine hall. For the first time, a black woman artist was selected for the event in 2019, and Kara Walker's exhibit *Fons Americanus* instantly sparked controversy. Walker crafted a 12-metre-high fountain that caricatured one of Britain's most famous public monuments. The Victoria Memorial was built outside the gates of the royal residence Buckingham Palace in 1911 to celebrate the achievements of the British Empire. At its top, a gilded Roman goddess—winged Victory—rises above Queen Victoria who is enthroned on a lower tier alongside her virtues of Truth, Justice and Motherhood. Bronze figures seated on lions guard each corner of the fountain's basin; one personifies "progress", two others manufacturing and agriculture, while one represents peace. To its passers-by, the memorial narrates how the British brought modern industry to the world under the reign of Victoria.

Walker's Fountain told another story. In the magazine *London Review of Books*, Cora Gilroy-Ware (2020: 16) argued that the installation was "deliberate profanation". At the top, a naked woman replaced the Roman goddess with her head thrown back. Water squirted out of her slit throat and nipples. A new set of virtues accompanied Queen Victoria: a slave-driving sea captain, a contorted man and a noose dangling from a tree. A sea of sharks and merchant ships filled the basin below. Upon reflection, the fountain created a bizarre counterpoint to the space in which it was displayed. The Tate Gallery, as the museum was originally known, was founded with an endowment from Sir Henry Tate. He was a sugar refiner who made his fortune from plantations in the British Caribbean. The institution harbouring Walker's fountain was itself a product of colonial trade and wealth amassed from an industry based on slave labour.

Fons Americanus offered a perspective on Britain's colonial history that is hidden from view in the Victoria Memorial and most other public art in the country. Walker's exhibition exposed the racialised and gendered violence inflicted throughout the empire to make "British" manufacturing possible. It also revealed the globalised networks delivering labour and resources to the imperial core, without which the products of industry and agriculture could ever exist. In this chapter, we explore the flows of labour, water, energy, land and emissions that enable the production and consumption of commodities today, developing a "virtual" political ecology perspective. Recent scholarship indicates that an asymmetric flow of resources from extractive peripheries to developed economic regions sustains the global economy. Building on discussions at the intersection of political ecology and the allied field of ecological economics, we

DOI: 10.4324/9781003095521-9

examine claims of ecological debt and argue that there is a moral obligation to act for environmental justice.

COMMODITIES AS MANIFESTATIONS OF GLOBAL FLOWS

If you belong to the global minority who can afford an iPhone, you own a device that quite likely has been assembled by workers in the city of Shenzhen, China. In 1982, Shenzhen had a population of about 350,000, but after it was declared a special economic zone during China's turn to a socialist market economy, rural migrants sought employment on the new assembly lines (Figure 9.1). In 2020, Shenzhen's population reached a colossal 17.6 million (Shenzhen Municipal Bureau of Statistics 2021). Without the labour of Shenzhen's workers, the world's electronics would remain an assortment of components with little practical or cultural value, yet with rare exceptions, the city's migrant workers remain invisible to those consuming the fruits of their labour. Ananya Roy (2011) describes an amusing exception that received a lot

FIGURE 9.1 Workers on the assembly line in Shenzhen, China, using UV light and epoxy glue to fit together electronics components

of attention online. When a British consumer known by his handle "markm49uk" unboxed a new iPhone, he discovered a photo had already been taken with the camera. A young female worker appeared on screen, busy assembling the phone while beaming a smile and making a peace sign. Writing on Shenzhen's role in the world economy, Roy argues that the normally anonymous Chinese worker was "asserting her place in the global value chain" (316). For a brief moment, the globalised networks that connect people with people, through things, and thus enable modern consumption, came into plain sight.

In an important text for economic geographers, David Harvey (1990) argues that as soon as we encounter a commodity—be it on the high street or the internet—we should exercise our "geographical imagination" to envision the social relations embedded in it. When I purchase a commodity, whose work am I buying? And under what conditions has that work taken place? A Marxist scholar, Harvey thinks through the notion of commodity fetishism (Chapter 3), which captures how modern consumers tend to see commodities as "things" with inherent value and purpose when their existence is inseparable from the work of others in a global commodity chain. And yet, Harvey acknowledges the difficulty of exercising a geographical imagination like this when the "spatial range of our own individual experience of procuring commodities in the marketplace bears no relationship to the spatial range over which the commodities themselves are produced" (423). The notion of commodity fetishism conceptualises how the products of human labour appear as objects in and of themselves, obscuring that labour, or as Marx (1976: 165) writes, how "the definite social relation between men [*sic*] … assumes … for them, the fantastic form of a relation between things".

Harvey's perspective is integral to scholarship in economic geography, but political ecologists have developed it further to show how not only exchanges of labour are manifest in commodities but also ecological goods. For instance, to keep an iPhone running, its battery must regularly be recharged. For this purpose, "markm49uk" would connect his phone to the UK electricity grid. Just as any machine, the phone exists as a continuation of an energy flow supplied to it through infrastructure, in keeping with the Second law of thermodynamics (Cederlöf 2021). In 2020, more than 35 per cent of all electricity in the UK grid was generated from natural gas. Thus, if we move from the phone, through its power cord and down the wire, we eventually run into the turbines in which this gas was combusted. The largest gas-turbine plant in the United Kingdom is located in Pembrokeshire, an agricultural region of Wales, which again testifies to the continuities that exist between urban consumption and rural livelihoods (Chapters 5 and 8). If we continue tracing the energy system, we next arrive in the North Sea. About two-thirds of the natural gas used in the British electricity sector is extracted there, in the UK's and Norway's oil fields. A further 10 per cent is imported in liquefied form from Qatar (BEIS 2021: chap. 4, 5). So, if we let our geographical imagination loose, where does the iPhone end in an ecological sense? In the palm of markm49uk's hand? In a turbine hall in Wales? In the North Sea or the Middle East?

Furthermore, the phone is made from a range of materials. Aluminium, iron and plastics are important, but also rarer metals, such as gold and tungsten. By accessing these materials, just as the energy stored in the battery (which is made from lithium; see Chapter 7), the phone user is situated in a vast global economy of resource flows. We can also take a broader view of the phone as a socio-ecological phenomenon. It is made up of various materials and energy, but what about the materials and the energy that have gone into the production of those goods in the first place? These are "virtual" resources that cannot be seen in the phone itself but without which the phone, its uses and cultural significances are unthinkable. A commodity, then, is not only a manifestation of the labour spent in manufacturing but also multiple direct and virtual ecological flows extending far beyond it in time and space.

In these terms, technological artefacts can be pictured as nodes in much wider networks of social and ecological relations. However, the labour and the resources that metabolise modern technological systems do not flow unintentionally across space but are mediated by markets, political institutions and everyday social relations. Reflecting on modern technology's enabling conditions, Alf Hornborg (2019) argues that consumers not only suffer from commodity fetishism but also "machine fetishism". We tend to see technological artefacts as autonomous objects—the inventions of human ingenuity—but just as a commodity is a product of labour, technology cannot be separated from the ecological flows it embodies. "The appearance of solid, material objectivity", Hornborg writes to define machine fetishism, "obscures the arbitrary social exchange relations that make the machine possible—for those who can afford it" (24).

EMBODIED LABOUR

It seems to be a deeply rooted proclivity in modern cultures to see objects as coextensive with their immediate boundaries. But this is not the case in all cultures (Ingold 2000), and it is a view that can be exposed to decolonial critique. Furthermore, to Europe's classical economists, the labour going into the production of a commodity was an essential feature of it. While their individual theories diverged, Adam Smith, David Ricardo and other nineteenth-century economists argued that value originates from the amount of labour required to produce a commodity. This labour was said to be "embodied" in the product. Karl Marx (1976) developed perhaps the most sophisticated labour theory of value. He argued that in any given historical-geographical context (characterised by the available technology and knowledge), the production of a commodity would on average require a certain amount of labour-time. Value was proportional to the socially necessary, average (or "abstract") labour-time embodied in a product.

At the heart of Marx's critique of capitalism was an observation about the production and distribution of value. Those producing commodities, he noted (i.e. the workers) received a wage in return for their labour-time. But since their employer (i.e. the capitalist) had to make a profit for the business not to fold, the workers could only be remunerated for a fraction of the value their labour produced. "Surplus value" was systematically transferred from the workers to the capitalist who profited off the work of others. And yet, since commodities tend to be fetishised in modern societies, the extraction of surplus value was culturally concealed through transactions of money. As David Graeber (2005: 429, original emphasis) writes, "Money represents the value of labor, but wage laborers work to get money; it thus becomes a representation that brings into being what it represents; it is therefore easy to see it as the source of that value, or *as* value (since again, from the laborer's perspective, it might as well be)".

As we discussed in Chapter 3, in the 1970s and 1980s neo-Marxist scholars sought to explain why the global economy was so unevenly developed—why the "developed" countries increased their levels of material consumption while the "underdeveloped" remained poor. A new generation of Latin American political economists developed a simple but powerful theory: international trade did not take place between equals but occurred within an unequal capitalist world system. A long history of colonialism had created a global core that was able to import a surplus of value from the periphery, thus perpetuating the development of the core and the underdevelopment of the periphery (Dos Santos 1978; Cardoso and Faletto 1979). Arghiri Emmanuel (1972) conceptualised this dependency relation in terms of "unequal exchange". First, he noted that the value of a commodity is derived from the labour embodied in it. When a person purchased a commodity, they also purchased the labour-time of someone else. If the person making the purchase was also a wage-labourer, they would trade their own labour-time for that of another. Second, populations in the developed countries had higher wages, with greater purchasing power, than populations in the underdeveloped world. This was an outcome of the industrial and agrarian economic structures that had developed in different parts of the world system in the colonial period. The differences in international wage levels, he concluded, permitted the developed countries to systematically import a surplus of value through commodity exchange, and the core thereby drained the periphery of development potential to its own benefit.

Today, labour theories of value are no longer in the economic mainstream. Not without good reason, neo-classical economists regard value as a matter of cultural preferences and consumer choices. However, debates on value do not disqualify the classical economists' more basic observation that commodities are manifestations of expended labour-time. If we accept that labour is a necessary requirement for bringing a commodity into existence, the commodity inevitably embodies labour-time, regardless of whether that labour is the source of value or not. When a consumer purchases a commodity, they enter into a social relation mediated by an object, and part of what they buy is the labour of others objectified in it. Nevertheless, as objects present themselves to our senses as bounded entities, the relations intersecting in them are concealed and easy to overlook.

VIRTUAL WATER

In the mid-1990s, a new idea revolutionised the study of water. A scholar of the Middle East, Tony Allan noted that water was incredibly scarce in the region, but even so, there were no major conflicts taking place over water. This was strange in an otherwise war-prone geopolitical environment. Allan argued that a trade of "virtual water" explained the paradox. The Jordan Basin (that is, the drainage area of the River Jordan) is politically subdivided into the states of Israel, Palestine and Jordan. To be self-sufficient, the three countries would need around 15 billion m^3 of water, yet there was less than 3 billion m^3 available to them in the basin. To grow one tonne of cereals on average requires 1,644 m^3 of water (Mekonnen and Hoekstra 2011: 1582), and such water-intensive agriculture would quickly drain the water budget (Table 9.1). Allan's important observation was that the three states recorded grain imports of more than 5 million tonnes annually. These cereals were grown in distant, water-rich regions, and by importing them, the Middle Eastern states alleviated the pressure on their own water resources by over 8 billion m^3 (5 million tonnes × 1,644 m^3). "[I]t is the

TABLE 9.1 Virtual water content of nine selected commodities

Commodity	Litres of Water Embodied in 1 kg	Visual Representation
Wheat	1,827	
Rice	2,497	
Lentils	5,874	
Cotton	9,982	
Coffee	18,900	
Chocolate	17,196	
Beef	15,415	
Pork	5,988	
Chicken	4,325	

To put the figures in context:

- If you brew a cup of coffee using 7 g of ground beans, you are using 135 L of embodied water
- If you wear a pair of jeans made from 800 g of cotton fabric, a staggering 8,000 L of water were needed in production

Source: Mekonnen and Hoekstra (2011, 2012)

global trading system that provides the strongest explanation for the water policies adopted by the Jordan Basin riparian states", Allan (2003: 143, emphasis added) concluded. "*Virtual water* enables serious water deficit economies to solve their water problems inexpensively, invisibly, and without political cost".

Over the past two decades, governments and businesses have embraced the idea of virtual water with remarkable alacrity. This is in part due to the managerial perspective informing discussions of the concept. Allan, for example, argues that international virtual water trade allows Middle Eastern countries to "construct false but widely accepted notions of water security" that reinforce "very sub-optimal water allocation policies" (143). If policymakers and markets account for virtual water, the argument goes, water deficits can be overcome with "optimised" imports of water-intensive products from other places.

Political ecologists have drawn attention to two dangers imbued with the market-based solutions towards which this argument gestures. Jessica Barnes (2013) notes that when water is regarded as a virtual commodity, traded on the world market, it is abstracted from its context of use. She shows how Egyptian government policies took account of virtual water and steered farmers away from rice farming. Seeing that water is scarce in Egypt, rice imports were deemed rational due to the virtual water embodied in the cereal. But Barnes questions whether "real" and "virtual" water really can be substituted in this manner. Drawing on her fieldwork, she explains how Egyptian farmers spoke of rice cultivation in terms of "washing" (*ghasal*) or "cleaning" (*nuzuf*) the land, and while some of the water supplied to the rice paddies percolated into the soil to become virtual water, the farmers regularly also grew rice with another purpose. By applying water to the paddies, they leached salts out of the soil, maintaining the long-term productivity of the land. In contrast to real water, Barnes argues, virtual water cannot leach salt. A notion of water that reduces H_2O to its flow through crops conceals the multiple roles played by water when it moves through lived-in landscapes.

Maria Beltrán and Esther Velázquez (2015) build on Barnes' analysis to draw attention to a second aspect shrouded by the notion of virtual water: the labour linked with farming and water management. Almería is a major agricultural export region in southern Spain, and as such, it is also an exporter of virtual water. Beltrán and Velázquez report that 40 per cent of the production costs linked with farming in Almería come from labour, and exports of virtual water, therefore, cannot be separated from this labour. The regional workforce is largely composed of migrant workers, and they work in a social context shaped by racialised tensions and a long history of organised class-struggle. In contrast to the way in which virtual water has been integrated with market-based solutions to improve water management, Beltrán and Velázquez call for a different interpretation of the concept. By trading virtual water, a financially strong actor can displace the social and environmental burden of its consumption in terms of labour and water use to a landscape located at a distance. To the extent that virtual water "flows" across borders, it cannot be separated from the political, social and economic mechanisms impelling it—mechanisms shaped by unequal power relations.

EMERGY

To develop a more complete virtual political ecology, we can continue considering the implications of embodied resources beyond labour and water. For example, we might analyse the levels of embodied CO_2 emissions or embodied timber in a product. The energy expenditures required to bring commodities to the market have received less attention in the literature, but this is a topic that has generated heated discussions among systems ecologists. Howard Odum was an important figure for the development of ecology as a natural science, and he was particularly interested in energy flows and the dynamics of self-organising systems. Odum introduced the notion of "emergy" (spelled with an *m*), which he at first defined as "embodied energy". However, the Second law of thermodynamics specifies that the energy used, directly or indirectly, to keep an organism alive or to manufacture a product cannot remain *in* the object since it dissipates into less useful forms when it is put to work. He, therefore, revised his characterisation of emergy to denote "energy memory" (Odum 1988).

A second problem with the definition of emergy was that when different quantities of energy were added up as part of a commodity's energy memory, the qualities of the expended energy were not accounted for. As noted in Chapter 7, one measure of wind and one measure of coal are not commensurate as they are qualitatively different phenomena. Odum reasoned that almost all energy on Earth originated from the sun, and he, therefore, used solar energy as a common denominator to assess the various energy transformations encapsulated in a product's emergy. While the standard unit for measuring energy is the joule, the "emjoule" became the unit for emergy values. One of Odum's most contested conclusions was that the ecosystems that embodied the greatest amounts of energy—demonstrating high emjoule values—tended to outcompete other ecosystems and were the most resilient to disturbance (Odum 1996). This "maximum power principle" explained the evolution of ecosystems, scaling up Darwin's theory of evolution, which had focused on the evolution of species.

Emergy analysis was at first reserved for nonhuman ecosystems, but in the 1980s, it was adapted to include human societies. Odum himself began relating emergy to money to examine the energetic basis of international trade. How many emjoules could one US dollar buy compared to one Bolivian boliviano (Odum 1996)? In *Underdeveloping the Amazon*, an important book for early Anglo-American political ecology, Stephen Bunker (1985) put Odum's concept to work and superimposed it on Emmanuel's theory of unequal exchange. Bunker argued that globally unequal wage-levels allowed the developed countries (or "articulated societies" in his terminology) to import a surplus of energy from "extractive economies" in the global periphery. The Amazon was underdeveloped because it was systematically drained of energy potential to the benefit of the core's development. "Thus", he wrote, "if society is composed of the energy it consumes, articulated societies consume extractive economies and their natural environments and in the process become more complex and more powerful" (34). For Bunker, dependency was a material relation, based on uneven energy flows.

In contrast to Emmanuel, Bunker stepped away from an idea of unequal exchange as a question of lopsided transfers of value extracted from labour and instead conceptualised it in terms of value extracted from nature. But his reasoning fell prey to an advanced form of environmental determinism. Determinist reasoning has a long history in attempts to explain social differences (Chapter 6). For instance, a long-lived determinist narrative holds that tropical environments are conducive to a particular kind of human behaviour, which is irrational and aggressive, in contrast to temperate climates that lead to level-headed, rational conduct (Arnold 2006). The main problem with determinist narratives—apart from having little support in evidence—is that they leave no space for other variables, including human agency, to explain social differences. In his study, Bunker explained social complexity as a direct outcome of the quantities of energy embodied in a social formation, and he argued that the developed countries systematically imported surpluses of energy to increase their complexity. While Bunker's argument can be criticised for its determinist slant, and energy may not be the source of value, it is nevertheless difficult to question the observations he made about the flow of energy on a global scale: energy continuously appears to move from extractive peripheries to economic cores.

EMBODIED LAND

To continue expanding the virtual political ecology perspective, it is interesting to consider the implications of embodied land. Picture an island for example. Islands are fascinating geographical formations to think with since, in an absolute sense, their shorelines demarcate a limited area available for agriculture, forestry, mining and other economic activities. However, to reproduce socio-economic relations, island populations tend to engage in extra-islandic trade, which allows them to rely on distant land uses and import land embodied in commodities. Reasoning by analogy, we can transfer the island logic to think about the land requirements of other geographical formations, such as cities—densely populated areas with limited land available for food production and other activities. In a recent Chinese study, Jieyu Wang et al. (2021) find that 80 per cent of the land needs of Shanghai, Tianjin and Beijing—three megacities—are satisfied through trade with other provinces. Writing on Phoenix, Arizona, Andrew Ross (2011: n.p.)

argues that "the desert metropolis necessarily depends on vast amounts of 'ghost acreage' elsewhere to sustain the daily needs of its population".[1] Commodities, then, not only embody distant labour, water and energy but also land uses.

In public debates, the idea of ecological footprints has garnered a lot of attention. Ecological footprints measure how large an area of eco-productive land a population requires to support its consumption. William Rees and Mathis Wackernagel (1996), who first proposed the idea, take paper as an example. They estimate that the average Canadian in the early 1990s used about 224 kg of paper products annually. One tonne of Canadian paper, in turn, required 1.8 m^3 of wood in production, and given a standardised average wood productivity rate used in ecological footprint assessments (2.3 m^3/ha/yr), Rees and Wackernagel conclude that the average Canadian required 0.19 ha of forestland in continuous production to support their paper consumption. Thus, newspapers and wrappers alike embody land, which can be traded as a virtual asset on the world market.

Ecological footprint analyses evaluate a population's environmental load relative to other populations in spatial terms. Developing their analysis, Rees and Wackernagel examine the metabolism of the city of Vancouver (paper products included), and just as Wang et al. did in the Chinese case, they find that the city relies on imports of embodied land on a scale vastly more extensive than the area of Vancouver "itself". "In effect", they write, "however healthy the region's economy appears to be in monetary terms, the Lower Fraser Basin [the area where Vancouver is located] is running a massive 'ecological deficit' with the rest of Canada and the world" (234). The distant land uses implicated in the daily running of the city thus appears to reinforce the theory of "planetary" or "extended urbanisation"—that the urban is now truly global in its influence (Chapter 5). Ecological footprint analysis has subsequently been refined, and assessments including food, housing, transportation, consumer goods and services show that the countries that tend to be ranked as the most "developed" in global comparisons—for example, in terms of the UNDP's Human Development Index (HDI; Figure 9.2)—also tend to demand much more land to sustain their consumption than what is available on the planet.

In part due to the methodological simplicity of the concept, ecological footprints demarcate what is known as "global hectares". The measure does not specify *what* land areas a population depends on, only that it relies on a certain amount of land "wherever on Earth that land is located" (Rees and Wackernagel 1996: 227). For this reason, political ecologists have taken a critical stance to ecological footprint analysis. If commodities embody land, *where* does that land exist? Who works it, and through what social, political and economic mechanisms are the virtual flows of land sustained? In an ambitious study, Luke Bergmann and Mollie Holmberg (2016) model the "tele-connections" that exist between consumers and spatially displaced land uses through the mediation of commodity chains and capital investments. They find that more than 50 per cent of all land in use globally services populations outside the country in which that land exists: the "median square meter", they report, is exported (933).

At Tate Modern, Kara Walker's fountain exposed how British manufacturing was closely interlaced with the colonial economy in the nineteenth century. In the article "Footprints in the Cotton Fields", Hornborg (2006) develops a method using historical statistics to assess the amounts of land and labour "saved" in British industry in 1850 through imports of cotton from overseas. He finds that the United Kingdom imported 254,921 tonnes of raw cotton in 1850, almost 90 per cent of which was grown on plantations in the American South. This

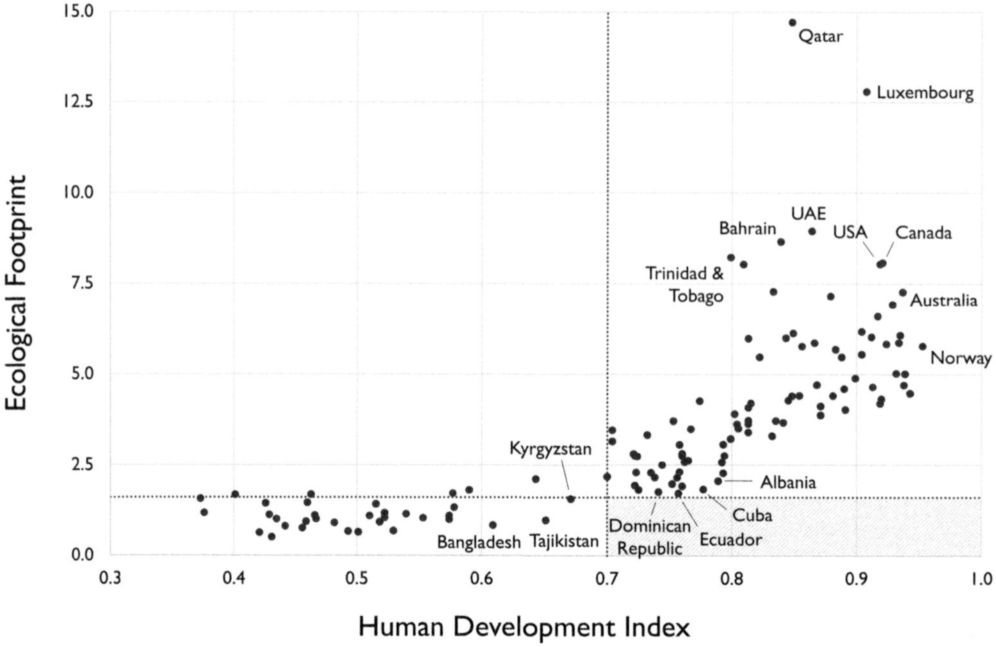

FIGURE 9.2 Sustainable development? Human Development Index and Ecological Footprint, 2017

The Human Development Index (HDI) is a composite measure of a population's life expectancy, education and per capita income. The Ecological Footprint measures how many global hectares (gha) per capita a population needs to sustain its average level of consumption. The United Nations defines a HDI over 0.7 as "high human development" while the planet currently can sustain an ecological footprint per capita up to 1.6 gha. To demonstrate "sustainable development" in these terms, a country must be placed in the shaded box in the graph's lower right-hand corner. No country in the available data set satisfied these conditions in 2017. At the same time, it is evident that the countries with the highest ecological footprints are some of the world's largest oil and gas producing countries, including Bahrain, Canada, Norway, Trinidad & Tobago, the United Arab Emirates, the United States and Qatar.

Source: York University Ecological Footprint Initiative & Global Footprint Network. National Footprint and Biocapacity Accounts, 2021 Edition. Data available at: data.footprintnetwork.org.

cotton had required 1.26 million ha of land and 702 million hours of mostly slave labour in production. Comparing the land and labour embodied in £1,000 worth of raw cotton with that required to produce £1,000 worth of cotton garments in Britain, he concludes that the imperial core gained a 210-per cent surplus of embodied land and a 123-per cent surplus of embodied labour through its textile industry in 1850. Reaching a more general conclusion, he argues that industrialisation, therefore, cannot be thought of only as a process of technological innovation: above all, industrialisation is enabled by a population's ability to appropriate space (land) and time (labour) from elsewhere.

In Chapter 8, we saw how a low-carbon energy transition gives rise to contested land-use changes. Photovoltaic solar panels, biofuel plantations and windfarms all take up space, and their relative land-intensities can be gauged in terms of their "power density" (Smil 2015). In a recent study of Cuban solar energy development, Hornborg et al. (2019) take a further step to also account for the virtual land uses needed to invest in photovoltaic systems. Interestingly, in Figure 9.2, Cuba emerges as one of the countries with the highest HDI scores and smallest ecological footprints in global comparison. The study finds that to produce only 3.6 per cent of Cuba's electricity using photovoltaics (which is the Cuban government's plan for 2030), 14,000 ha must be covered with solar panels directly. However, to manufacture the panels, Cuba also needs to access more than 440,000 ha embodied in the solar panels, most of which is land imported from other countries and China in particular. This means that for Cuba to pursue a low-carbon transition, it must build local solar parks but also displace the environmental load of its investments to lands beyond the Cuban islands. The ability to transition to a solar-powered economy, then, is dependent on a population's position in the global political economy.

EXPANDING THE SYSTEM BOUNDARIES: ECOLOGICALLY UNEQUAL EXCHANGE

In her landmark *For Space*, the geographer Doreen Massey (2005) advocates for a relational understanding of space. She argues that a particular geographical imagination developed within the history of modernity and writes:

> One characteristic of this [imagination] was an assumption of isomorphism between space/place on the one hand and society/culture on the other. … [It] was firmly established that space and society mapped on to each other and that together they were, in some sense "from the beginning", divided up. … "Places" came to be seen as bounded, with their own internally generated authenticities, and defined by their difference from other places which lay outside, beyond their borders.
>
> (64)

The anthropologist Eric Wolf (2010) also famously argues for a relational understanding of the world economy. Using a powerful metaphor, he writes:

> By turning names into things we create false models of reality. By endowing nations, societies, or cultures with the qualities of internally homogenous and externally bounded objects, we create a model of the world as a global pool hall in which the entities spin off each other like so many hard and round billiard balls. Thus it becomes easy to sort the world into differently colored balls, to declare that "East is East, and West is West, and never the twain shall meet".
>
> (6–7)

Massey and Wolf both criticise the kind of "spatial fetishism" we discussed in relation to the key concept of scale in Chapter 3 in which the separation of the world into nations and cultures—a mosaic of internally coherent spaces—blocks out the processes that connect and constitute those spaces in the first place. Instead, as Wolf writes, it is necessary "to think of human aggregates as 'inextricably involved with other aggregates, near and far, in weblike, netlike connections'" (385). A virtual political ecology perspective similarly invites us to transcend the appearance of things as internally coherent entities and instead ask, in what ways do commodities represent weblike, netlike connections?

To think relationally about "things", ecologists speak about the need to define the "system boundaries". A system boundary delimits everything that is included in the phenomenon under study from that which is not. For example, with narrowly drawn system boundaries around an iPhone, the phone is only constituted by the materials and the energy present in it at a given moment. But with more extensive system boundaries, the materials, energy, water and labour necessary to produce and charge it are also included. System boundaries is an epistemological concept, because to gain knowledge about a commodity, it is necessary to first define its system boundaries—to define where it begins and ends. Crucially, political ecologists have demonstrated that the definition of system boundaries is far from an innocent task: it is well-known that differently drawn system boundaries can lead to entirely different assessments of socio-environmental impact (Newell and Vos 2011; Guibrunet et al. 2017).

Sweden is often put forward as a country at the forefront of a low-carbon transition. Swedish government statistics show that, between 1990 and 2019, the emissions of greenhouse gases decreased by 29 per cent in the country (Statistics Sweden 2021). This figure is based on emissions from production on national territory, where transportation together with the steel, cement and oil-refining industries are the major emitters. Nevertheless, the statistics database allows us to expand the system boundaries and include emissions from fuels used in international shipping and air travel to and from Sweden. With the system boundaries drawn in this more extensive way, it turns out that the emissions only decreased by 19 per cent—more than a third less than the original figure.

In 2019, the Swedish government decided that so-called consumption-based emissions should be accounted for in national reports. These are emissions that result from consumption within national boundaries, even though the commodities may have been produced abroad. Virtual emissions, in other words. Aiming to develop a policy for reducing consumption-based emissions, the government reported that when the system boundaries were redrawn in this fashion, the emissions from commodities produced abroad exceeded those recorded nationally with a remarkable 70 per cent (Regeringen 2020: 8). In an in-depth study of "displaced" Swedish emissions, Cindy Isenhour and Kuishuang Feng (2016: 322) find that when consumption-based emissions are accounted for, Swedish emissions in fact increased by 25 per cent between 1990 and 2010. The bulk of those emissions occurred in China.

So, which is the correct way of drawing the system boundaries? The only available answer is that it is an epistemological requirement to draw boundaries around phenomena in order to gain knowledge about them. As Donna Haraway (1991: 201) argues, "Objects are boundary projects. What boundaries provisionally contain remains generative, productive of meanings and bodies". In comparison, Gustav Cederlöf and Hornborg (2021) argue that system boundaries are inherently cultural phenomena as they reflect ontological convictions that resonate with the interests of particular social groups: "If system boundaries are seen to be cultural products", they write, "it follows that they are integral to sustaining different power-serving worldviews"

(120). For example, in the case of production-based emissions, which does not account for virtually traded emissions, a country can register a success in emissions reductions by displacing the environmental load of consumption to another areal unit (cf. Cousins and Newell 2015).

In an epistemological sense, a virtual political ecology perspective implies that the system boundaries should be expanded as fully as possible around commodities to bring out the flows of labour and resources embodied in them. A group of political ecologists and ecological economists has recently designed their research following this ideal to assess how materials flow through the world's commodity chains. An uneven geographical pattern emerges in their research. Sweden, as we have seen, has reduced its carbon emissions from domestic production in recent years, but meantime, increasing national consumption levels have resulted in emissions in China instead (Isenhour and Feng 2016). In the nineteenth century, the British textile industry expanded at the same time as space and time were appropriated in colonised areas of the British Empire (Hornborg 2006). More broadly, low-income countries exporting to relatively wealthier countries systematically experience higher levels of deforestation, and low-income and export dependence, therefore, seem to correlate with increasing environmental degradation (Jorgensen 2006). One of the most ambitious studies to date concludes that between 1990 and 2015, the world's high-income countries all imported surpluses of virtual resources while the middle- and low-income countries were net exporters (Dorninger et al. 2021). Reviewing the literature, Jessica Givens et al. (2019: 2) summarise the evidence thus: "Global South nations are structurally positioned as both a tap for resources and a sink for waste within the world-economic system of extraction, production, and consumption".

The systematic transfer of ecological goods from extractive peripheries to core regions constitute "ecologically unequal exchange". Introducing this term, Hornborg (1998) built on a world-systems perspective, drawing inspiration from work on embodied labour and emergy in which "unequal exchange" was theorised in terms of asymmetric transfers of "value" (Emmanuel 1972; Bunker 1985). An anthropologist, Hornborg instead maintains that value is determined by cultural preferences. And yet, cultural beliefs and economic activities cannot be separated from the material flows with which they are intertwined. The production of value is predicated on a dissipation of energy and raw materials, and to sustain this entropic process (Chapter 7), world-economic cores extract labour, land and resources from global peripheries even as they return waste to them. The occurrence of ecologically unequal exchange, then, is at odds with the tendency in modern societies to objectify technology, regarding it as an autonomous domain. Nevertheless, such machine fetishism plays an important ideological role in the global economy: when the products of socio-ecological exchange relations appear in the form of autonomous objects—as iPhones and solar panels—ecologically unequal exchange is culturally concealed. Machine fetishism mystifies and thereby reinforces the political economic interests served by the globally asymmetric flows of resources.

FROM ECOLOGICAL DEBT TO ENVIRONMENTAL JUSTICE

Many of the key concepts we have introduced in this book—power, scale, nature and urbanisation to mention a few—are tools with which we can analyse how economic activities and environmental changes are co-productive of social inequalities. The theory of ecologically unequal exchange is also an attempt to make sense of the patterns that characterise the global economy in terms of material flows. A theory like this identifies gross social inequalities, but even so, it is not in itself a normative theory. However, it is not a far step to argue that the inequalities we observe at the same time are injustices: that inequalities are morally wrong. Indeed, many political ecologists are guided by a sense of moral urgency, motivating them in their work to create a more socially and environmentally just world. Writing from a Latin American vantage point, Enrique Leff (2015: 47) notably positions political ecology as "a field of political ethics"; Lucy Jarosz (2004: 917) goes one step further and defines political ecology "as ethical practice".

Discussions of ecologically unequal exchange are twinned with the normative concept of ecological debt. This idea first emerged from a social movement advocating the cancellation of financial debt in the 1990s. While governments in the global north insisted on the repayment of financial debt, the Ecuadorian NGO Acción Ecológica drew attention to the ecological debts amassed by the global north through virtual resource trade. Reflecting on the concept, Joan Martinez-Alier et al. (2014) argue that "the ecological debt arises from the plunder of resources (ecologically unequal trade) and also from the occupation by rich countries of disproportionate "environmental space" ... to dump excessive amounts of carbon dioxide in the oceans and the atmosphere which belong to all humans equally" (28). The existence of ecological debt raises questions of historical accountability: to what extent must current generations in the most developed countries be held accountable for socially produced natural phenomena, such as climate change and environmental degradation? The questions are delicate, because as Rees and Wackernagel (1996) argue, global sustainability cannot be financed by ecological deficits. "[S]imple physics", they write, "dictates that not all countries or regions can be net importers of biophysical capacity" (235).

The issue of ecological debt resonates with calls for environmental justice. Environmental justice first became the motto for Afro-American communities struggling against toxic waste dumping in the United States (Bullard 1990). Coupled with broader calls for climate justice, it is today a watchword for a global social movement (Sultana 2022). Scholarship on environmental justice makes three important distinctions that help guide action in this direction. The first distinction highlights issues of distribution. *Distributional justice* is achieved when access to resources is no longer socially differentiated and vulnerable communities are no longer excessively exposed to risk. If ecologically unequal exchange theorises an asymmetric transfer of ecological goods on a global scale, ecological debt can be understood as a principle of distributional justice (Warlenius et al. 2015). The second distinction foregrounds recognition and the silencing of marginalised voices in environmental struggles. *Recognitional justice* is achieved when cultural differences, marginalised voices and non-Western epistemologies are acknowledged in public debate and policymaking. The final distinction focuses on representation. Beyond recognising difference, *representational justice* asserts the need for inclusion in decision-making processes and more democratic environmental governance (Fraser 2008; Schlosberg 2007).

Environmental justice struggles tend to flare up at the resource frontier when extractive processes lead to dispossession, marginalisation, degradation and elevated risk. To make these struggles visible, political ecologists associated with the Autonomous University of Barcelona curate the *Global Atlas of Environmental Justice* (ejatlas.org) (Figure 9.3). While indigenous peoples represent about 5 per cent of the world population, indigenous communities are involved in a staggering 40 per cent of the cases documented in the atlas (Temper et al. 2018). Allied with indigenous social movements, Lina Álvarez and Brendan Coolsaet (2018) have recently pointed

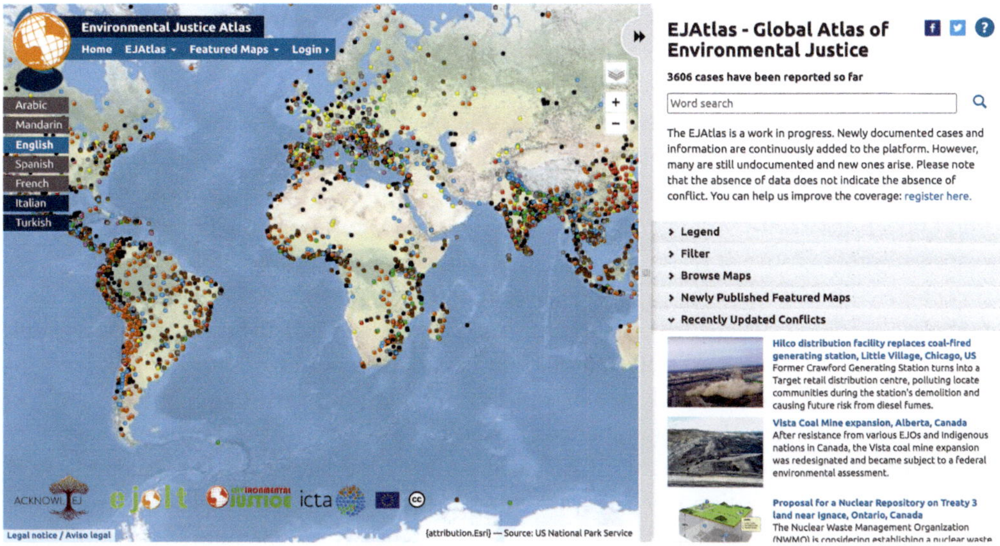

FIGURE 9.3 Map from the EJAtlas: Global Atlas of Environmental Justice

The EJAtlas, which is a work in progress, currently documents over 3,600 environmental conflicts. The conflicts concern a wide range of issues, including biodiversity conservation, energy production, mineral extraction, nuclear waste treatment and water management.

Source: ejatlas.org; Temper et al. (2015)

out that the environmental justice discourse, despite its good intentions, needs to be decolonised. Álvarez and Coolsaet echo Leah Temper (2016: 98) who argues that liberal notions of justice universalise a Western worldview. These notions, she writes, "are founded on 'universal access' to rights grounded in a western knowledge system based on 'objective' knowledge". When indigenous communities are recognised and represented in settler-colonial institutions, they are expected to address their claims in institutions they object to—colonial institutions in which non-Western epistemologies at best are subsumed. Decolonial environmental justice is instead tantamount to indigenous communities' self-government and territorial sovereignty.

In a field that takes the study of power relations and the silencing of marginalised voices as its points of departure, it is often difficult, on a personal level, to avoid asking what *my* role is in all of this. What *I* can do. Political ecologists are sometimes criticised for being great at exposing injustices but poor at offering alternative solutions. While participation in environmental justice struggles clearly offers a corrective to this critique, the argument is generating discussions about the implications of an "affirmative political ecology" (Alhojärvi and Sirviö 2018). How can the methods and the insights developed in the field be used for change? This, of course, speaks to an understanding of political ecology as praxis. Action can imply getting engaged in local community work and social movements, in politics, scholarship or education. Reflecting on her work in Latin America, Susan Paulson (2018: 87) argues that political ecologists urgently need to move beyond critique "toward practicing change in one's own position and supporting synergy among tangible pathways studied in contexts around the world". She also emphasises the potential that exists in engaging with heterodox future-looking perspectives, such as epistemologies of the south, decolonial praxis and degrowth, many of which we have discussed in this book (Escobar 2016; Demaria et al. 2019; Kothari et al. 2019).

Meantime, political ecologists ingrained in the Anglo-American tradition who feel compelled to confront environmental injustices often find inspiration in a famous passage from Marx's *Theses on Feuerbach*. In the eleventh thesis, Marx (1845, original emphases) writes that "Philosophers have hitherto only *interpreted* the world in various ways; the point it to *change* it". If we encounter injustice, it is also our duty to act on it. In Chapter 1, we noted how the political ecology community has branched out over the past decades to encompass a diverse range of scholars, activists, teachers, artists, journalists and policymakers. Change is afoot when people (and increasingly, nonhumans) with multifarious experiences of environmental injustices gain voice in the field, even as such change cannot be separated from hard-fought efforts of giving voice (Bryant 2015). It is the responsibility of established scholars and policymakers to use their institutional clout to make space for marginalised voices. As we have argued, it is incumbent on those in a privileged position to co-produce knowledge *with* and *from* subaltern perspectives.

There is, of course, no one best way of acting for change. It is sometimes said that the most radical action in a warming world is to plant a tree—making a claim to space, nature and the future. Needless to say, that act raises questions as to whose land the tree grows on; who owns the tree; under what conditions it is managed and through whose labour; how access to its fruits, shade and other social benefits is regulated; and not least importantly, through whose narratives and symbolic systems the tree attains meaning. In exploring human-environment interactions, activism is the primary motivation for many political ecologists to ask such questions, using their research to the benefit of social movements and environmental justice organisations. Others are guided by a desire to make their research policy-relevant, hoping to develop more inclusive institutions for environmental governance. For others still, it is the opportunity of refining the questions asked, developing the methods used and improving the conceptual frameworks put to work to make sense of the world that is the main motivation behind their work. Stirred by intellectual curiosity, they find teaching, writing and learning to be the principal means of working for change. Having written this book, we hope that it has provoked you to ask critical questions about the interplay between environmental change and social inequality, and we hope that it kindles you to act for a more just sustainable future.

CONCLUSION AND LEARNING OUTCOMES

At Tate Modern, Kara Walker gave voice to experiences that normally are silenced in Western accounts of the colonial past and the globalised present. Her installation drew attention to the enabling conditions of modern industry that tend to be hidden from view. The concepts of commodity fetishism and machine fetishism capture how modern consumers tend to regard objects as autonomous "things", obscuring the socio-ecological exchange relations of which they are manifestations. To the contrary, a virtual political ecology perspective defines the widest possible system boundaries to account for the labour, water, energy, land and emissions embodied in commodities. Operationalising this definition and using methods from ecological economics, a growing literature now indicates that ecologically unequal exchange characterises the globalised economy. The asymmetric transfers of labour and ecological goods on a global scale exhibit great inequalities in distribution and raise questions about ecological debt and differentiated historical responsibilities for climate change. When faced with environmental injustices, we are convinced of the moral obligation to act with marginalised communities for change.

Having read this chapter, we hope that you can:

- Explain the concepts of commodity fetishism and machine fetishism as well as their ideological role in modern societies
- Account for the virtual political ecology concepts of embodied labour, virtual water, emergy and ecological footprints
- Analyse how the system boundaries are drawn around a commodity or other object in assessments of social and environmental impact
- Explain the theories of ecologically unequal exchange and ecological debt
- Take an informed stance on environmental justice and reflect on how you might work for a more just sustainable world

FURTHER READING

Givens, Jennifer E., Xiaorui Huang, and Andrew K. Jorgensen. 2019. "Ecologically Unequal Exchange: A Theory of Global Environmental *Injustice*." *Sociology Compass* 13(5): e12693.

This article is a great text to read after this chapter if you want to gain a more in-depth understanding of the theory of ecologically unequal exchange and the literature that is forming around it. It is particularly interesting for the way it connects ecologically unequal exchange with scholarship on environmental justice.

Martinez-Alier, Joan, et al. 2014. "Between Activism and Science: Grassroots Concepts for Sustainability Coined by Environmental Justice Organizations." *Journal of Political Ecology* 21: 19–60.

The literature on environmental justice (which variously also is framed in terms of climate justice, water justice and energy justice) is expansive, but this paper does an excellent job in defining a set of core concepts that each has emerged through the work of environmental justice organisations.

NOTE

1. The concept of "ghost acreage" was first developed by the neo-Malthusian scientist Georg Borgström (1965: 71), who defined it as "the computed, non-visible acreage which a country would require as a supplement to its present visible agricultural acreage in the form of tilled land in order to be able to feed itself".

BIBLIOGRAPHY

Alhojärvi, Tuomo, and Heikki Sirviö. 2018. "Affirming Political Ecology: Seeds, Hatchets and Situated Entanglements." *Nordia Geographical Publications* 47(5): 1–6.

Allan, J. A. 2003. "Virtual Water Eliminates Water Wars? A Case Study from the Middle East." In *Virtual Water Trade: Proceedings of the International Expert Meeting on Virtual Water Trade*, edited by A.Y. Hoekstra, 137–145. Value of Water Research Report Series No. 12. Delft, Netherlands: IHE Delft.

Álvarez, Lina, and Brendan Coolsaet. 2018. "Decolonizing Environmental Justice Studies: A Latin American Perspective." *Capitalism Nature Socialism* 31(2): 50–69.

Arnold, David. 2006. *The Tropics and the Traveling Gaze: India, Landscape, and Science, 1800–1856*. Seattle, WA: University of Washington Press.

Barnes, Jessica. 2013. "Water, Water Everywhere but Not a Drop to Drink: The False Promise of Virtual Water." *Critique of Anthropology* 33(4): 371–389.

BEIS. 2021. "Digest of UK Energy Statistics: Annual Data for UK, 2020". Statistical release, UK Department for Business, Energy & Industrial Strategy. https://www.gov.uk/government/statistics/digest-of-uk-energy-statistics-dukes-2021. Accessed: 10 January 2022.

Beltrán, Maria J., and Esther Velázquez. 2015. "The Political Ecology of Virtual Water in Southern Spain." *International Journal of Urban and Regional Research* 39(5): 1020–1036.

Bergmann, Luke, and Mollie Holmberg. 2016. "Land in Motion." *Annals of the American Association of Geographers* 106(4): 932–956.

Borgström, Georg. 1965. *The Hungry Planet: The Modern World at the Edge of Famine*. New York, NY: Macmillan.

Bryant, Raymond L. 2015. "Reflecting on Political Ecology." In *The International Handbook of Political Ecology*, edited by Raymond L. Bryant, 14–24. Cheltenham, UK: Edward Elgar.

Bullard, Robert D. 1990. *Dumping in Dixie: Race, Class, and Environmental Quality*. Boulder, CO: Westview Press.

Bunker, Stephen G. 1985. *Underdeveloping the Amazon: Extraction, Unequal Exchange, and the Failure of the Modern State*. Urbana, IL: University of Illinois Press.

Cardoso, Fernando Enrique, and Enzo Faletto. 1979. *Dependency and Development in Latin America*. Translated by Marjory Mattingly Urquidi. Berkeley, CA: University of California Press.

Cederlöf, Gustav. 2021. "Out of Steam: Energy, Materiality, and Political Ecology." *Progress in Human Geography* 45(1): 70–87.

Cederlöf, Gustav, and Alf Hornborg. 2021. "System Boundaries as Epistemological and Ethnographic Problems: Assessing Energy Technology and Socio-Environmental Impact." *Journal of Political Ecology* 28: 111–123.

Cousins, Joshua J., and Joshua P. Newell. 2015. "A Political-Industrial Ecology of Water Supply Infrastructure for Los Angeles." *Geoforum* 58: 38–50.

Demaria, Federico, Giorgos Kallis, and Karen Bakker. 2019. "Geographies of Degrowth: Nowtopias, Resurgences and the Decolonization of Imaginaries and Places." *Environment and Planning E: Nature and Space* 2(3): 431–450.

Dorninger, Christian, Alf Hornborg, David J. Abson, Henrik con Wehrden, Anke Schaffartzik, Stefan Giljum, John-Oliver Engler, Robert L. Feller, Klaus Hubacek, and Hanspeter Wieland. 2021. "Global Patterns of Ecologically Unequal Exchange: Implications for Sustainability in the 21st Century." *Ecological Economics* 179: 106824.

Dos Santos, Theotonio. 1978. *Imperialismo y dependencia* [*Imperialism and Dependency*]. Mexico City, Mexico: Edición Era.

Emmanuel, Arghiri. 1972. *Unequal Exchange: A Study of the Imperialism of Trade*. Translated by Brian Pearce. New York, NY: Monthly Review Press.

Escobar, Arturo. 2016. "Thinking-Feeling with the Earth: Territorial Struggles and the Ontological Dimension of the Epistemologies of the South." *AIBR. Revista de Antropología Iberoamericana* 11(1): 11–32.

Fraser, Nancy. 2008. "Abnormal Justice." *Critical Inquiry* 34(3): 393–422.

Gilroy-Ware, Cora. 2020. "At Tate Modern." *London Review of Books* 42(3): 16.

Givens, Jennifer E., Xiaorui Huang, and Andrew K. Jorgensen. 2019. "Ecologically Unequal Exchange: A Theory of Global Environmental *Injustice*." *Sociology Compass* 13(5): e12693.

Graeber, David. 2005. "Fetishism as Social Creativity: Or, Fetishes Are Gods in the Process of Construction." *Anthropological Theory* 5(4): 407–438.

Guibrunet, Louise, Martin Sanzana Calvet, and Vanesa Castán Broto. 2017. "Flows, System Boundaries and the Politics of Urban Metabolism: Waste Management in Mexico City and Santiago de Chile." *Geoforum* 85: 353–367.

Haraway, Donna J. 1991. *Simians, Cyborgs, and Women: The Reinvention of Nature*. London, UK: Free Association Books.

Harvey, David. 1990. "Between Space and Time: Reflections on the Geographical Imagination." *Annals of the Association of American Geographers* 80(3): 418–434.

Hornborg, Alf. 1998. "Towards an Ecological Theory of Unequal Exchange: Articulating World System Theory and Ecological Economics." *Ecological Economics* 25(1): 127–136.

Hornborg, Alf. 2006. "Footprints in the Cotton Fields: The Industrial Revolution as Time-Space Appropriation and Environmental Load Displacement." *Ecological Economics* 59(1): 74–81.

Hornborg, Alf. 2019. *Nature, Society, and Justice in the Anthropocene: Unraveling the Money-Energy-Technology Complex*. Cambridge, UK: Cambridge University Press.

Hornborg, Alf, Gustav Cederlöf, and Andreas Roos. 2019. "Has Cuba Exposed the Myth of 'Free' Solar Power? Energy, Space, and Justice." *Environment and Planning E: Nature and Space* 2(4): 989–1008.

Ingold, Tim. 2000. *The Perception of the Environment: Essays on Livelihood, Dwelling and Skill*. London, UK: Routledge.

Isenhour, Cindy, and Kuishuang Feng. 2016. "Decoupling and Displaced Emissions: On Swedish Consumers, Chinese Producers and Policy to Address the Climate Impact of Consumption." *Journal of Cleaner Production* 134A: 320–329.

Jarosz, Lucy. 2004. "Political Ecology as Ethical Practice." *Political Geography* 23(7): 917–927.

Jorgensen, Andrew K. 2006. "Unequal Ecological Exchange and Environmental Destruction: A Theoretical Proposition and Cross-National Study of Deforestation, 1990–2000." *Rural Sociology* 71(4): 685–712.

Kothari, Ashish, Ariel Salleh, Arturo Escobar, Federico Demaria, and Alberto Acosta, eds. 2019. *Pluriverse: A Post-Development Dictionary*. New Delhi, India: Tulika Books.

Leff, Enrique. 2015. "Political Ecology: A Latin American Perspective." *Desenvolvimento e Meio Ambiente* 35: 29–64.

Martinez-Alier, Joan, et al. 2014. "Between Activism and Science: Grassroots Concepts for Sustainability Coined by Environmental Justice Organizations." *Journal of Political Ecology* 21: 19–60.

Marx, Karl. 1845. *Theses of Feuerbach*. https://www.marxists.org/archive/marx/works/1845/theses/index.htm. Accessed: 5 January 2022.

Marx, Karl. 1976 (1867). *Capital: A Critique of Political Economy*, vol. 1. Translated by Ben Fowkes. Harmondsworth, UK: Penguin.

Massey, Doreen. 2005. *For Space*. London, UK: Sage.

Mekonnen, M. M., and A. Y. Hoekstra. 2011. "The Green, Blue and Grey Water Footprint of Crops and Derived Crop Products." *Hydrology and Earth System Sciences* 15: 1577–1600.

Mekonnen, Mesfin M., and Arjen Y. Hoekstra. 2012. "A Global Assessment of the Water Footprint of Farm Animal Products." *Ecosystems* 15(3): 401–415.

Newell, Joshua P., and Robert O. Vos. 2011. "'Papering' over Space and Place: Product Carbon Footprint Modeling in the Global Paper Industry." *Annals of the Association of American Geographers* 101(4): 730–741.

Odum, Howard T. 1988. "Self-Organization, Transformity, and Information." *Science* 242(4882): 1132–1139.

Odum, Howard T. 1996. *Environmental Accounting: Emergy and Environmental Decision Making*. New York, NY: John Wiley & Sons.

Paulson, Susan. 2018. "Pluriversal Learning: Pathways toward a World of Many Worlds." *Nordia Geographical Perspectives* 47(5): 85–109.

Rees, William, and Mathis Wackernagel. 1996. "Urban Ecological Footprints: Why Cities Cannot Be Sustainable—and Why They Are a Key to Sustainability." *Environmental Impact Assessment Review* 16: 223–248.

Regeringen. 2020. "Tilläggsdirektiv till Miljömålsberedningen (M 2010:04)—strategi för minskad klimatpåverkan från konsumtion" [Additional Directive for the Cross-Party Committee on Environmental Objectives (M 2010:04)—Strategy for Reduced Climate Impact from Consumption]. Government of Sweden, Committee Directive, Dir. 2020:110. Stockholm, Sweden: Ministry of the Environment.

Ross, Andrew. 2011. "Bird on Fire: Lessons from the World's Least Sustainable City." *Places Journal*, November.

Roy, Ananya. 2011. "Postcolonial Urbanism: Speed, Hysteria, Mass Dreams." In *Worlding Cities: Asian Experiments and the Art of Being Global*, edited by Ananya Roy and Aihwa Ong, 307–355. Malden, MA: Wiley-Blackwell.

Schlosberg, David. 2007. *Defining Environmental Justice: Theories, Movements, and Nature*. Oxford, UK: Oxford University Press.

Shenzhen Municipal Bureau of Statistics. 2021."深圳市第七次全国人口普查公报[1] (第一号)" [Communiqué of the Seventh National Census of Shenzhen City[1] (No. 1)]. http://tjj.sz.gov.cn/zwgk/zfxxgkml/tjsj/tjgb/content/post_8771927.html. Accessed: 26 November 2021.

Smil, Vaclav. 2015. *Power Density: A Key to Understanding Energy Sources and Uses*. Cambridge, MA: MIT Press.

Statistics Sweden. 2021. "Total Emissions and Removals of Greenhouse Gases by Greenhouse Gas and Sector. Year 1990–2020." Statistical database, https://www.statistikdatabasen.scb.se/pxweb/en/ssd/START__MI__MI0107/TotaltUtslappN. Accessed: 5 January 2022.

Sultana, Farhana. 2022. "Critical Climate Justice." *The Geographical Journal* 188(1): 118–124.

Temper, Leah. 2016. "Blocking Pipelines, Unsettling Environmental Justice: From Rights to Nature to Responsibility to Territory." *Local Environment* 24(2): 94–112.

Temper, Leah, Daniela del Bene, and Joan Martínez Alier. 2015. "Mapping the Frontiers and Front Lines of Global Environmental Justice: The EJAtlas." *Journal of Political Ecology* 22: 255–278.

Temper, Leah, Federico Demaria, Arnim Scheidel, Daniela del Bene, and Joan Martinez-Alier. 2018. "The Global Environmental Justice Atlas (EJAtlas): Ecological Distribution Conflicts as Forces for Sustainability." *Sustainability Science* 13: 573–584.

Wang, Jieyu, Shaojian Wang, and Chunshan Zhou. 2021. "Quantifying Embodied Cultivated Land-Use Change and Its Socioeconomic Driving Forces in China." *Applied Geography* 137: 102601.

Warlenius, Rikard, Gregory Pierce, and Vasna Ramasar. 2015. "Reversing the Arrow of Arrears: The Concept of 'Ecological Debt' and Its Value for Environmental Justice." *Global Environmental Change* 30: 21–30.

Wolf, Eric R. 2010. *Europe and the People without History*, 2nd ed. Oakland, CA: University of California Press.

Activities

Seminar activity 1: Discovering political ecology

Aim: To discover political ecology as a community of practice/praxis

Learning outcomes:

- Identify some key concerns of different political ecologists
- Apply a critical understanding of political ecology as epistemology, ontology and praxis

Task: Interpretation of journal abstracts

Paul Robbins (2020) defines political ecology as "a community of practice", suggesting that the field is defined by the questions, methods and theories posed and developed by a community of scholars. Recently, Tracey Osborne (2017) also emphasised the synergies existing between theory and practice, arguing that political ecology is best defined as "a community of praxis". In this activity, you will gain a sense of the work different political ecologists are currently embarking on. The exercise also invites you to distinguish between the uses of "political ecology" as epistemology, ontology and praxis.

STEP ONE: READ THREE SUMMARIES OF RECENT WORK IN THE FIELD

The *Journal of Political Ecology* (*JPE*) is the longest-established journal in the field and is very unusual in being entirely free to both readers and authors. Most journals either charge your university libraries vast subscription fees or charge authors for publishing in them. While we have offered you our understanding of political ecology as a community in Chapter 1, it is also important to see how others define the field. *JPE* is a good source for doing this.

Progress in Human Geography is a very different kind of journal and is very much part of the commercial side of academic publishing. Over the last 25 years, it has published some of

DOI: 10.4324/9781003095521-10

the most important summaries of political ecology in its "Progress reports". These are basically updates on the field and provide snapshots of what a single academic (whose perspective, remember, is situated and partial) considers to be important at that time.

1 Visit the "About" page of *JPE* and read about what kind of papers the journal seeks to publish: https://journals.librarypublishing.arizona.edu/jpe/site/About.
2 Now, choose at random one of the issues published in the last five years and pick an article that looks interesting to you. Read the abstract for this paper.
3 Next, choose one "Progress report" on Political Ecology or Urban Political Ecology published in *Progress in Human Geography*. Again, read the abstract for this report.

STEP TWO: SUMMARISE THE KEY CONCERNS OF YOUR THREE SOURCES

1 Take ten minutes to jot down what you think are (1) the key concerns of *JPE*; (2) the aims of the paper in *JPE*; and (3) the key themes identified within the Progress report.
2 Working with one of your peers, try to explain what you think the authors are doing in each of these pieces.
3 Finally, repeat the exercise, focusing on epistemology, ontology and praxis. Where are each of the authors making epistemological and ontological claims? In what circumstances do they view political ecology as a form of praxis?

REFERENCES CITED

Osborne, Tracey. 2017. "Public Political Ecology: A Community of Praxis for Earth Stewardship." *Journal of Political Ecology* 24: 843–860.

Robbins, Paul. 2020. *Political Ecology: A Critical Introduction*, 3rd ed. Oxford, UK: Wiley-Blackwell.

Seminar activity 2: Power

Aim: To interrogate different forms of power

Learning outcomes:

- Account for power as a resource, social relation, ideology, discourse and knowledge

Task: Preparing an action plan

This activity is based around a scenario ideally to be discussed as part of a group.

SCENARIO: A CENTRE FOR SUSTAINABLE DEVELOPMENT?

Your university has just accepted a multi-million-dollar donation from a large fossil fuel company in order to establish a Centre for Sustainable Development. For the last century, this fossil fuel company has made very large profits from the extraction of oil and natural gas in different parts of the world. Over the last thirty years, the company and some of its subsidiaries have come under particular scrutiny for a series of major oil spills on the lands of peasant farmers in a West African country. A social movement has formed seeking to challenge the environmental degradation of these farmlands, but a military dictatorship running the country has suppressed it. One of the activists involved has tragically been killed.

While the vast majority of the company's profits are still made from the extraction and sale of fossil fuels, it now claims to be wanting to move beyond its core business: it has expanded into renewable energy production, solar and wind production in particular. The Centre for Sustainable Development is part of this move, and the purpose of it is to fund scholarships for students alongside new posts in teaching and research for academic staff. Around the world, university budgets are coming under increasing pressure following a series of cuts to its funding streams, and the university administration is keen to accept the donation. Some students, however, are furious that it would even consider doing so.

THE TASK

The scenario described above is ripe for political ecological analysis. An author such as Michael Watts (2004) has also written about oil and power in the Niger Delta, while Gabriela Valdivia (2018) has written incisive analyses of the politics of oil extraction in Ecuador. In this activity, you should work individually or in a group to develop an action plan for the university as well as for the social movement that is likely to oppose the investment.

1. Identify the different forms of power at work in the scenario context. Try to use the typology provided in Chapter 2 and feel free to speculate a little beyond the brief summary given above with the aid of Watts' and Valdivia's research.

2 Next, put these perspectives on power to work to consider how they might be addressed: (a) from the perspective of the university administration, and (b) from the perspective of a student social movement committed to environmental justice to develop an action plan for each actor.

REFERENCES CITED

Valdivia, Gabriela. 2018. "'Wagering Life' in the Petro-City: Embodied Ecologies of Oil Flow, Capitalism, and Justice in Esmeraldas, Ecuador." *Annals of the American Association of Geographers* 108(2): 549–557.

Watts, Michael. 2004. "Resource Curse? Governmentality, Oil and Power in the Niger Delta, Nigeria." *Geopolitics* 9(1): 50–80.

Seminar activity 3: Scale

Aim: To explore the ways in which scale can be used in political ecology

Learning outcomes:

- Explain why it is always necessary to make choices about the scales at which a study is carried out and explanation takes place
- Observe patterns of socio-ecological change in the world around you and begin to see how they cohere around specific temporal and spatial scales
- Account for the notion of "scale jumping" and its role as a political tactic for the environmental justice movement

Task: Group discussion

In a research project called the Plumbing Poverty Project (PPP), a group of political ecologists sought to understand who does and who does not have access to piped water in the United States. In a very surprising finding, they found that half a million households across the country lacked indoor plumbing. Staggering numbers of people, in other words, were found to be without piped water and/or access to a flushing toilet in their homes. Even more surprisingly, some of the wealthiest cities in the United States had among the highest proportions of households in this situation of "plumbing poverty". And yet, it was notable that people of colour to a significantly larger extent were exposed to this form of environmental injustice.

The PPP's findings have been published in two leading academic journals (Deitz and Meehan 2019; Meehan et al. 2020) and also in mainstream news outlets such as *The Guardian* (Lakhani et al. 2021) and *National Public Radio*. The research exposes not only historic and emerging forms of environmental injustice but also the politics and power relations producing these forms of environmental injustice. In this sense, the research is quintessentially political and ecological. Less discussed in the work of the PPP, however, is the scalar production of this environmental injustice and the ways in which the interpretation of the qualitative and quantitative data they collected can be aided by a scalar imagination. This is the focus on this activity.

First, read the excellent summary of the PPP team's research by Nina Lakhani et al. (2021) in *The Guardian*: https://www.theguardian.com/us-news/2021/sep/27/water-almost-half-million-us-households-lack-indoor-plumbing.

Second, first individually or in a pair, and then with your peers, discuss the following questions:

1. What choices around scale has the PPP team done to analyse plumbing poverty? Could a different choice have led to a contrasting result or a sharpened understanding of the phenomenon?
2. How have different, material, economic, social and political patterns converged in particular locations in the United States to produce "plumbing poverty"?
3. How might social movements organise around different scales to challenge the environmental injustice of plumbing poverty? At what scale is action most likely to be effective and is there a need for the movements to "jump scale"?

REFERENCES CITED

Deitz, Shiloh, and Katie Meehan. 2019. "Plumbing Poverty: Mapping Hot Spots of Racial and Geographic Inequality in US Household Water Insecurity." *Annals of the American Association of Geographers* 109(4): 1092–1109.

Lakhani, Nina, Maanvi Singh, and Rashida Kamal. 2021. "Almost Half a Million US Households Lack Indoor Plumbing: 'The Conditions Are Inhumane'." *The Guardian*, 27 September. https://www.theguardian.com/us-news/2021/sep/27/water-almost-half-million-us-households-lack-indoor-plumbing. Accessed: 29 September 2022.

Meehan, Katie, Jason R. Jurjevich, Nicholas M. J. W. Chun, and Justin Sherrill. 2020. "Geographies of Insecure Water Access and the Housing-Water Nexus in US Cities." *Proceedings of the National Academy of Sciences* 117(46): 28700–28707.

Seminar activity 4: Nature

Aim: To critically reflect on the idea of "wilderness" and the political implications of it

Learning outcomes:

- Interrogate epistemological frameworks that conceptualise nature and "wilderness" as distinct from society
- Explain the ideological function of claims of a "natural order"

Task: Group discussion

The broadcaster Sir David Attenborough is often celebrated as one of the world's foremost makers of wildlife and natural history documentaries. Originally trained in zoology and geology at the University of Cambridge in the 1940s, he released the miniseries *Dynasties* in 2018 in which "Five extraordinary animals, each in a heroic struggle against rivals and against the forces of nature, fight[...] for their own survival and for the future of their dynasties" (BBC n.d.). Deeply sceptical about the depiction of wilderness in the series—an environment in which humans are completely absent—the environmental journalist George Monbiot (2018) wrote a scathing critique for the newspaper *The Guardian*:

> For many years, wildlife film-making has presented a pristine living world. It has created an impression of security and abundance, even in places afflicted by cascading ecological collapse. The cameras reassure us that there are vast tracts of wilderness in which wildlife continues to thrive. They cultivate complacency, not action.
>
> You cannot do such a thing passively. Wildlife film-makers I know tell me that the effort to portray what looks like an untouched ecosystem becomes harder every year. They have to choose their camera angles ever more carefully to exclude the evidence of destruction, travel further to find the Edens they depict. They know—and many feel deeply uncomfortable about it—that they are telling a false story, creating a fairytale world that persuades us all is well, in the midst of an existential crisis.

Having read Chapter 4, work through the following questions and then discuss them with your peers:

1. Why does Monbiot argue that it is problematic to portray animals "in a heroic struggle ... against the forces of nature"? Why would political ecologists argue that this is a false or at least very partial representation of animal life?
2. In Chapter 2, we outlined "ideology" as a form of power. Building on Monbiot's critique, how might the *Dynasties* series be seen to reproduce an "ideology of nature" and what interests would that ideology serve to upkeep?
3. National parks are often portrayed as spaces "set aside" by humans to provide "sanctuaries" for wildlife. Many studies have shown how this prevailing form of biodiversity

conservation often results in the dispossession of poor communities who live off the land where the parks are built (e.g. Kabra and Das 2022). Beyond this troubling political dynamic, the idea of humans managing "wild spaces" seems like a paradox in itself. Is it possible to create wilderness? If not, how might endangered species be saved from extinction?

4 Political ecologists usually focus their research on humans and human relations with the environment. By failing to include humans, Monbiot argues that traditional wildlife documentaries "cultivate complacency" to questions of species extinction and ecological disturbance. However, to the extent that they present animals as sentient subjects, how might documentaries such as *Dynasties* change the ways we think about "power", which normally focuses on human processes and practices?

REFERENCES CITED

BBC. n.d. "Dynasties." BBC Earth website, https://www.bbcearth.com/shows/dynasties. Accessed: 14 September 2022.

Kabra, Asmita, and Budhaditya Das. 2022. "Aye for the Tiger: Hegemony, Authority, and Volition in India's Regime of Dispossession for Conservation." *Oxford Development Studies* 50(1): 44–61.

Monbiot, George. 2018. "David Attenborough Has Betrayed the Living World He Loves." *The Guardian*, 7 November. https://www.theguardian.com/commentisfree/2018/nov/07/david-attenborough-world-environment-bbc-films. Accessed: 14 September 2022.

Seminar activity 5: Urbanisation

Aim: To gain a situated understanding of how urbanisation connects cities with places beyond them

Learning outcome:

- Interpret the city and its hinterlands as a process of urbanisation

Task: Photography and group discussion

Two signal concepts for urban political ecology are "circulations and metabolisms". These point to an understanding of the city as a meeting place of relationships, flows and practices and of urbanisation as a continuously ongoing process. In this activity, you will start from the area where you live to explore this understanding of the city. The activity unfolds in two steps, the first to be undertaken individually or in a pair and the second in a group.

STEP ONE: DOCUMENT AN URBAN OBJECT

The first step is for you, and every seminar participant, to take a photo of an object in the city where you live that somehow is part of an urban flow (e.g. a flow of food, water, energy, waste, sewage, information, transport, etc.). If you live in the countryside, you can think of the flow in reverse—photograph an object that connects your area with a city.

1. Paste your photo into a text document and give it a short title.
2. In no more than 300 words, carefully describe the object. Insofar it is relevant for your object, think about its materials, form, function, smell, sound, feel, taste, rhythms and how people (and animals) behave around it while you are observing it. Think about how you are delimiting your object (where does it begin and end?).

STEP TWO: GROUP DISCUSSION

1. Share your objects and descriptions of them.
2. Discuss together what actors the objects affect, for example gendered/racialised/classed social groups, businesses, the state/municipality, NGOs, workers, farmers, animals, etc.
3. Discuss in what ways the objects are part of economic and ecological flows that connect the city to its hinterlands.
4. Discuss what the objects say about social inequality and environmental unsustainability but also the possibility of creating a more just sustainable city.

Seminar activity 6: Water

Aim: To explore one aspect of the hydrosocial cycle, namely how the flow of water in rivers is intertwined with larger political economies

Learning outcomes:

- Apply the hydrosocial cycle as a political ecological concept
- Account for the changing political economies through which networked water supplies have been provided

Task: Interpretation of maps and aerial photos

Luleälven (River Lule) is Sweden's second largest river by volume, flowing from the Sulitelma massif on the Swedish-Norwegian border into the Gulf of Bothnia. In Lule Sami, the river is known as *Julevädno*, which means "the river of the Forest Sami". The area around *Julevädno* has been inhabited by the indigenous Sami people for generations, but in the 1950s, the Swedish state started to regulate the river to develop the country's hydroelectric capacity. Integrating the river in the political economy of the welfare state, the damming led to the forced displacement of many Sami families as their settlements, fishing sites and reindeer pastures were flooded. Today, the state-owned utility Vattenfall operates 15 dams along the river, controlling its flow, and the annual production amounts to about 10 per cent of Sweden's total electricity output.

In a text titled "Being May-Britt Öhman: Or, Reflections on My Own Colonized Mind Regarding Hydropower Constructions in Sápmi", the Sami historian May-Britt Öhman (2010: 18) writes of her relation to the river:

> I grew up at a time when the hydroelectric scheme was already a *fait accompli*. Born in 1966, I have never known of anything else. Growing up I never realized that we went fishing on a reservoir, that is, in a state controlled lake. When I, as a child, never could catch a fish in the river I always thought, it was because I was a bad angler. My own experience of Swedish industrialization is rather one of the idea of progress—the fascination of the large hydropower plants—and the production of electricity. That is the story that I have learnt, in school and through the promotion of hydropower as a green electricity alternative. Also, my father had an engineering firm, a subcontractor to the iron works in Luleå, providing electrical installations. It was only by talking to A. [Öhman's uncle], I began to sense and understand another aspect of the modernization story, the loss and destruction of both habitat and economy.

In this activity, you will examine how Vattenfall's construction of one dam, intervening in the hydrosocial cycle, resulted in large-scale environmental change and produced a new kind of nature around—and including—River Lule. To do this, you will use a mapping tool provided by the Swedish National Land Survey (Lantmäteriet) available at https://minkarta.lantmateriet.se. Unless you read Swedish, begin by selecting "English" under the "Languages" tab.

In the top right-hand corner, you can find an icon with a green-blue-and-red square-shaped symbol called "Change map layer". Click it and you will find a range of different cartographic datasets:

- **Map** is a topographic map of Sweden, showing elevation, placenames as well as other geospatial information such as roads and reindeer fences (the latter are black lines with double dashes on them).
- **Aerial photo** displays images of the ground taken from an aeroplane. Each pixel in the image represents 0.5 m^2 of the surface of Sweden. The images in this map layer are the most current available.
- **Aerial photo ca. 1960** shows images taken at some point between 1949 and 1970 (the surface of Sweden is too large to be photographed in one year, so images from several years have been combined).

Using the "Search" window in the top-right-hand corner, now search for "Heliga fallet", which means "The Holy Waterfall" in Swedish. This takes you to Seitevare, which is the northernmost dam in Lesser Lule River—the river's name before it joins with Greater Lule River at Vuollerim. In the "map" and "aerial photo" layers, you will be able to see the dam wall, which is about 1.6 km long and 106 m high. In "aerial photo ca. 1960" you can see the dam under construction. The construction of Seitevare started in 1962 and the dam was inaugurated in 1968.

1 Compare the aerial photos from today and ca. 1960 and, in writing, describe the land cover and environmental changes you can discern as a result from the construction works.

   In the "Menu" on the right (the icon marked with three black parallel lines), you can find a measuring tool allowing you to measure distances (in metres) and areas (in hectares). Use this tool to gather data for your descriptions so that your interpretation is well-supported with evidence.

   Discuss your findings as a group.

2 Use the "map" layer to get a sense of land use in the area today. Can you draw any conclusions about how the dam construction changed the land use patterns in the area?

   The name of *Heliga fallet* (or *Bassegårttje* in Lule Sami) indicates that this place has particular cultural and spiritual significances that go beyond the current use for electricity production. Indeed, before the dam was built, there was a waterfall, 50 m wide and 7 m high, downstream of the dam. A large rock in the shape of male face protruded from the waterfall. This was a *seite*—a Sami cult object—at which the white reindeer was sacrificed to the gods. Studying the aerial imagery, can you locate the former waterfall and describe how the land use around it has changed as the area was integrated in the political economy of the Swedish welfare state?

   Discuss your findings as a group.

REFERENCES CITED

Öhman, May-Britt. 2010. "Being May-Britt Öhman: Or, Reflections on My Own Colonized Mind Regarding Hydropower Constructions in Sápmi." In *Travelling Thoughtfulness: Feminist Technoscience Stories*, edited by Pirjo Elovaara, Johanna Sefyrin, May-Britt Öhman, and Christina Björkman, 260–283. Umeå, Sweden: Umeå University Department of Informatics.

Seminar activity 7: Energy

Aim: To explore, first, how electricity is embedded in your own everyday life, and second, how your electricity use, in turn, connects you with politicised environments in which energy is extracted

Learning outcomes:

- Assess how particular energy forms are embedded in landscapes of consumption, where they serve social, economic, cultural and political interests
- Analyse how energy extraction takes place along a commodity chain, linking environmental actors in an often-global network with socially and economically unequal outcomes

Task: Home inventory and online research

This activity unfolds in two steps, which are carried out individually. You can then compare and discuss your results with your peers.

STEP ONE: EMBEDDED ENERGY

The first step is to identify how electricity is embedded in your household practices. Copy the table below and make an inventory of all the electrical appliances you use—directly or indirectly—where you live.

Appliance name	Power rating (W)	Time of use (h)	Energy use (Wh)
Ex: Coffee maker	*1,080*	*0.5*	*540*
...	...	...	...
		Total use (Wh):	
		Total use (kWh):	

1 *List all the appliances in the first column.* Think carefully about lighting, space heating/cooling, cooking, refrigeration, water heating and communications (phones, computers, TVs, transportation, etc.).

2 *Find the appliances' power ratings.* Almost all appliances have their power ratings printed on a label: you are looking for the rating stated in watts (W). If you cannot find it, search for it online.

3 *Estimate the amount of time you use the appliance in a day.* Note the time to the nearest quarter; for example, if you use your desk light for 5¾ h per day, record "5.75".

4 Multiply the power rating with the time of use (W × h) and record this in the "Energy use (Wh)" column.

5 Sum up your use in the "Total use (Wh)" box. Divide the result by 1,000 to convert it to the standard unit kilowatt-hour (kWh).

You will now have a sense of how electricity is embedded in your everyday life. How do different appliances compare with each other? Does any result surprise you?

STEP TWO: EXTRACTED ENERGY

The second step is to research what extracted energy resources you depend on via your electricity use, moving across the commodity chain from consumption to production.

1 Searching online, find out what the "energy mix" looks like in the country where you live (i.e. how much coal, how much nuclear, how much wind, etc.). The International Energy Agency's country-based data (http://www.iea.org/countries) can be a useful website to explore for this information.
2 Using the above information, identify how many kWh you consume based on each energy source. For example, if your daily estimated consumption is 5 kWh, and electricity is generated from 50 per cent coal and 50 per cent wind in your country, your result will be 2.5 kWh coal, 2.5 kWh wind.
3 A key interest for political ecologists is how energy extraction gives rise to environmental justice conflicts. In Chapter 9, we introduce the *Global Atlas of Environmental Justice* (https://ejatlas.org), which is a mapping tool that documents environmental justice conflicts around the world. Use the atlas to explore how the extraction of the energy sources you depend on causes conflict.

Seminar activity 8: Fields & forests

Aim: To explore key questions associated with land ownership, land management and agricultural development

Learning outcome:

- Interpret longstanding debates on commons and enclosures, as well as the economic interests and lived experiences they express

Task: Preparing a project brief

This activity is based around a scenario inspired by a development project in India in which one of us worked in the early 2010s. Read through the scenario, continue to research the context and then develop responses to the three questions, individually or in a small group. Finally, discuss the conclusions drawn by different groups, all groups together.

SCENARIO: LAND MANAGEMENT FOR A WOMEN'S COLLECTIVE

The Deccan Plateau stretches across southwestern India. It is a high, flat terrain wedged between mountain ranges, and it extends from the city of Pune in the west to Hyderabad in the east and Bengaluru in the south. The Deccan is the heartland of India's cotton industry, and most cotton here is of the "Bt" variety. A product of the Green Revolution, Bt seeds have been genetically modified to produce an insecticide known as *Bacillus thuringiensis* (Bt), which kills bollworm. Bt cotton is exclusively sourced from a multinational corporation, which also supplies synthetic pesticides and fertilisers required for a successful harvest.

While many large landowners profit from the industry, the levels of social inequality are high in the Deccan. Farming requires seeds, fertilisers and pesticides to be bought at the beginning of the crop cycle, and the multinational corporation offers the farmers credits, allowing them to pay for inputs after the harvest. The threat of crop failure puts many farmers at risk. Many poor women work as day labourers on the cotton plantations, and they supplement their small income with rice rationed by the state to those living below the poverty line.

Working with a local NGO, a group of women have formed a *sangham*—a women's collective. Each *sangham* member saves ₹5 a week, and when she reaches ₹100, she gets a loan of ₹2,000 from the NGO to invest in a small plot of land. Today, the women have recovered 30 ha of land and cultivate it with millets for self-subsistence.

THE TASK

The *sangham* is discussing how to manage their land most effectively, and they are seeking your advice on the following questions in writing:

1. What would the benefits be of managing the land as a commons? How might they regulate access to it?
2. What would the benefits be of enclosing the land for individual plots to be managed as private property?
3. How will their choices affect their relations with the NGO, the multinational corporation, large landowners and other actors?

Seminar activity 9: Virtual political ecology

Aim: To explore work for environmental justice that goes on in your local area

Learning outcome:

- Take an informed stance on environmental justice and reflect on how you might work for a more just sustainable world

Task: Interview and group discussion

Questions of environmental justice are all around us, whether directly through a local struggle or in relation to inequalities shaped by local consumption patterns, national politics and global environmental change. In this activity, you will explore how people are working for environmental justice in the area where you live, work or go to university.

STEP ONE: IDENTIFY AND INTERVIEW A LOCAL ENVIRONMENTAL JUSTICE ORGANISATION

First, individually or in a pair, you should research what environmental justice organisations that exist in your local area. Use your contacts or search the internet. You can also identify a member of parliament or a municipal representative who takes an interest in environmental issues. Next, contact them to ask for an interview in which you ask about their thoughts on and work for environmental justice:

1. What are the most burning environmental issues for them? Are these also justice-related issues?
2. How do they understand the connection between local experiences and global processes?
3. How do they work, if at all, for environmental justice?
4. What are their main achievements so far and what are their main challenges going forward?

STEP TWO: GROUP DISCUSSION

Second, compare and discuss your findings with your peers:

1. How do the ways in which the people you have interviewed speak about environmental issues and environmental justice differ?
2. What can the key concepts of power, scale, nature, urbanisation and a "virtual" political ecology perspective tell you about their ideas and work?
3. How could their analyses and advocacy become more effective?
4. Would you want to get involved with their work and why (not)?

Index

Note: *Italicised* and **bold** pages refer to figures and tables, respectively.

For Product Safety Concerns and Information please contact our EU representative GPSR@taylorandfrancis.com Taylor & Francis Verlag GmbH, Kaufingerstraße 24, 80331 München, Germany